Vedic Mathematics

The Problem Solver

Vedic Mathematics

The Problem Solver

maple press

VEDIC MATHEMATICS THE PROBLEM SOLVER

Published by

MAPLE PRESS PRIVATE LIMITED
office: A-63, Sector 58, Noida 201301, U.P., India
phone: +91 120 455 3581, 455 3583
email: info@maplepress.co.in
website: www.maplepress.co.in

Printed in 2024, India

ISBN: 978-93-50332-59-7

Contents

1

Introduction to Vedic Mathematics

The Sanskrit word *Veda* is derived from the root *Vid,* meaning to 'know without limit' and it covers all *Veda Sakhas* known to humanity. The *Veda* is a storehouse of all knowledge and ever enlightening as it is delved deeper. Vedic wisdom is lively in the consciousness of living saints. It is also well preserved in the Vedic literature. The inner evidence of the available Vedic knowledge of the available Vedic literature makes it out that originally the whole range of the pure knowledge i.e., Vedic knowledge was vibrating from a single wholesome Veda.

This knowledge was organized by Maharishi Ved Vyas as four *Vedas,* namely, *Rig Ved, Yajura Ved, Sama Ved* and *Atharva Veda.* The knowledge of *Rig Veda* admitted 21 branches while the knowledge of other three *Vedas* namely, *Yajur Veda,* Sama Veda *and Atharva Veda* respectively admitted 101,1000 and 9 branches. The organizational format of each Vedic branch consisted of four folds designated as *Samhita, Brahmana, Aryanak* and *Upanishad.* As such 21 + 101 + 1000 + 9 = 1131 Vedic branches had 1131 *Samhita?,,* 1131 *Brahmanas,* 1131 *Arynaks* and 1131 *Upanishads.* These 1131x4 = 4524 scriptures together came to be known as Vedic wisdom. In addition corresponding to each Ved, as applied value of the pure knowledge of the *Ved* is *Upved.* The four *Upvedas* are *Ayur Veda, Dhanura Veda, Gandharva Veda* and *Sthapatya Veda.* Our present day Mathematics, science, and technology come within the range of *Sthapatya Veda. Manasara* is one such scripture of *Sthapatya Veda.*

Some of opinions on Vedic Mathematics are as follows:

(a) Mathematics, derived from the *Veda,* provides one line, mental and super fast methods along with quick cross-checking systems.

(b) Mathematics converts tedious maths into a playful and blissful subject, which students learn with smiles.

(c) Vedic Mathematics offers a new and entirely different approach to Mathematics based on pattern recognition. It allows for constant expression of a student's creativity and is found to be easier to learn.

(d) In this system, for any problem, there is always one general technique applicable to all cases, and also a number of special pattern problems. The element of choice and flexibility at each stage keeps the mind lively and alert to develop clarity of mind and intuition, and thereby a holistic development of the human brain automatically takes place.

(e) Vedic Mathematics with its special features has the inbuilt potential to solve the problem *of maths-anxiety.*

The experience of teaching Vedic Mathematics methods to children has shown that a high degree of mathematical ability. It can be attained from an early stage while the subject is enjoyed for its own merits. In fact, the Vedic Mathematics system is 'one of the most delightful chapters of the 20th century mathematical history'. The system has great educational value because the *sutras* contain techniques for performing some elementary mathematical operations in simple ways, and results are obtained quickly. Vedic Mathematics can be used to remove mathphobia and can be taught to (school) children as improvement material along with other high-speed methods.

Vedic Mathematics is easier to learn, faster to use and less prone to error than usual methods. Besides, the techniques of Vedic Maths not only enable the students to solve specific mathematical problems; they also develop creativity, logical thinking and intuition.

The real four and higher spaces formats of Vedic comprehensions are new wonderful worlds of very rich Mathematics, which may ensure us powerful technologies, and much possible disciplines of knowledge. The basic understanding pointed out is the way the cosmic surface constitutes and binds the solid granules as synthetic solids manifesting in the cosmos. The work of some mathematicians has added a new dimension to the dialogue initiated with the interpretation of the *Ganita Sutras.*

Rig Ved Samhita is the first Vedic scripture. It is the first book of the mankind. The Mathematics precedes the composition *of Rig Ved Samhita. Vedic* Mathematics helped to transform the universal set of knowledge as a speaking language and in the process it itself as well transformed as such and assimilated its identity into the *Vedas.*

2

Ganita Sutras

Ganita Sutras is a complete system of Vedic maxims. They are 29 in number and 16 of them are designated as *Sutras* and other 13 as *Upsutras.* Many people have been convinced about the potentialities of the *Sutras* when the working rules of these *Sutras* are demonstrated. Since then many scholars have further demonstrated about their potentialities to make the Mathematics much easier and more effectively within the comprehensions of much larger population and also about their utility and academic values in many ways. Some of the *ganita sutras* are discussed below:

EKADHIKENA PURVENA 1

There are three technical terms, which are fundamentals to the working domain of the *Sutra.* They are: (1) *Eka* (one), (2) *Edhikena* (more than), and (3) *Purvena* (previous). The first concept is the wholesome concept and it manifests in many ways like a unit, unit entity, unit measure, the counting number 1, the close interval of unit length etc. The second concept 'more than' is the concept of increase, the concept of comparison of one entity being bigger than / more than other entity. The third concept is the concept of 'order'. It is the concept of a queue, a pair of entities of a queue of which one is previous as comparison to the second and the second automatically being subsequent to the first. Therefore, the *Sutra* 1 has many applied values which would work out for us (1) counting, (2) arrangement of counting pebbles, (3) arrangement of beads along a thread, (4) the queue of points along with line etc.

If we try to chase the repeated applications of 'one more than the previous one', we can have countably infinite steps. Not only that, we can by a suitable choice of'one' and the application of the working rule *of Sutra* 1, have a chase for uncountable infinity as the line is by having sequential placements of points as counts with index values allotted as per the working rule *of Sutra* 1 from the set of natural numbers.

GANITA UPSUTRA 1

In *Upsutra* 1, we find a pair of technical words: *(I) Ami* (to follow), and (2) *Rupyena* (the form of). When 'one' is viewed as synthetic i.e., composed of parts, then each of its parts would be symmetrical to the whole and it would be proportionate of the whole. As such, (1) part as symmetrical to the whole, and (2) part as proportionate of the whole would be two distinct aspects of the same organisation.

The applied values of *Upsutra* would help us reach at unit interval as replica of infinite line, the universality of *Yathapinde tatha brahmande* (as in the body so in the universe) and *Yatha brahmande tatha pinde* (as in the universe so in the body) as well as the universality of the *Upanishadic* comprehensions as that even when full is taken out of full, there still remains full. With this, when one would work, *Sutra* 1 with double of a unit instead of a unit or with any multiple of a unit instead of a unit, one would see that the rule works. With this, the *Sutra* 1 coupled with *Upsutra* 1 makes the potentialities of *Sutra* 1 manifold.

GANITA SUTRA 2

The four technical words in *Ganita Sutra 2* are: (1) *Nikhilam* (all), (2) *Navastasca* (from nine), (3) *Caramam* (last), and (4) *Dastah* (ten). Nine numerals and ten-place value system are the two basic steps in terms of which the infinitely long line stands tamed. The counting pebbles technique requires infinite pebbles for coverage of infinitely long line. However, symmetry/ proportions/ replica rules suggest that proper system can be constructed to express 'infinity in terms of finite'.

Vedic seers comprehended ten-place value system. From the method of expression tenth counting pebble as 10 as in two digit form, we can notice that here lies a shift from single digit expression for first 9 counts to the next multiple digit expression initiation for expression of tenth count onwards. We also know that first 9 counts as well can be expressed in double digit forms as: 01, 02, 03, 04, 05, 06, 07, 08 and 09. Therefore, the first thing about this transition is the introduction of the tenth symbol for the zero count as '0'. Further, it also would be interesting to note that the double digit form for 1 as 01 provides us an expression for the tenth count as 10 simply by having swapping of places for first and the second digit. Further, the above interlink of 01 and 10 takes us to their link and interlocking as reflection pair of numbers as much as that the formats beneath 01 and 10 get interconnected like image of an arrow through looking mirror gets interlocked with the arrow itself. The placements of arrow, mirror, and the image of the arrow work out a di-monad format with mirror playing the role of joint of the di-monad. This as such is a big leap forward from the organisation format *of Sutra* 1. This had become possible with tile transition stage of *Upsutra 1*. Therefore, *Sutra 1, Upsutra 1* and *Sutra* 2 together make much interesting dimensions to the Mathematics.

The above organisation *of Sutra* 1, *Upsutra* 1 and *Sutra* 2 takes us to the middle of interval. It is this way unique fixation of the middle makes the Mathematics workable in terms of 'half unit', like the organisation in terms of' half dimension'. With this, it becomes possible to go to a number which exists prior to 'one' i.e., half. It is a big achievement over the Mathematics of 'one' without predecessor. However, simultaneously, it makes the working very delicate, as we shall be permitting the second half to remain dormant while structures because of the second half also shall be silently marking their presence. We cannot afford to ignore the existence of the structures because of the second dormant half. If one is to see the type of difficulties in which the Mathematics when worked without taking care for the second dormant half, one is simply to pose to oneself as to why from hypercircle-8 onwards the values start decreasing. It simply happens as the linear order worked in terms of half-dimension helps us sequentially increase only until seventh step parallel to the possible seven geometries of 3-space of linear order.

The organisation of pairing of two halves with the help of a mirror at the middle is unique organisation of many features, of which the most important is the feature of common remainder for both the halves while both halves are divided by a common divisor. Though this property may appear to be obvious but in fact when the same is viewed in the context of uncountably infinite points being handled in terms of counting numbers, one would realise, how important this property of constant remainders emerges to be. This is precisely *Upsutra 2.*

GANITA UPSUTRA 2

The basic / technical word availed by the text of *Upsutra 2* is *'Sesa'* (remainder). The definition of remainder is that which remains at the end. The basic operation is the operation of'division'. This *Upsutra* along with *Upsutra I* makes the Mathematics of *Upsutras* a very interesting branch in itself. The one which permitted itself as a pair of halves, sequentially by the rule of symmetry of *Upsutra* 1 and the rule of equality of remainders of division of equals takes the process of partitioning of 'one' into as many equal parts as one would like to rjave. This makes the external expansion of counts from one to two, two to three and so on parallel to the internal partitioning from wholesome one to a pair of parts, then to three parts, four parts and so on. With this, the arrays of mathematical tools become sufficiently large to work out very rich structures only in terms of 'parts', of which the working with 'half is most rewarding.

URDHVA -TIRYAGBHYAM

The basic / technical words availed by the text are: (1) *Urdhva* (vertical) and (2) *Tiryag* (crosswise). It is one of the working rules of this *Sutra,* which would help us to reach at its organisational format as well. Let us multiply 11 by 11. We shall be getting the product equal to 121. Let us have a close look at this number. The central digit of this number is 2. It admits expression as 1+1. Let us replace the central digit 2 with its above expression 1 + 1. The three parts of the number 121 would permit depiction as: 1, 1 + 1, 1. Now, let us chase the multiplication the *Sutra* way: The first *Sutra* gives us the working rule to proceed step by step as 'one more than the previous one'. When we are working only with 'half, therefore, in this case of 121 we shall be requiring the last two digits 21 or the first two digits 12. Therefore, when we shall be having 1 as start with position, as a next step under the rule *of Sutra* 1, we shall be reaching at 12. Then the reflection image of 12 would give us 21 and we shall be obtaining 121. If we would have been multiplying 111*111, we would be obtaining in three steps 123 and its image 321 which when combined would give us 12321 as the product of 111x111. Likewise if we would have been multiplying 1111x1111, we would be obtaining in four steps 1234 and image 4321 which when combined would give us the required product as 1234321. The process can be extended. The product 11x11 being the product of pair of numbers of double digits, therefore, these four digits, here each being 1, can be expressed as the numbers located at the four corners of a square. The pair of above corners may be taken as representing the first number 11 and the pair of lower corners may be taken as second number 11. Now, when we shall be multiplying the first digit of lower number 11 with the pair of digits of the above number 11 one by one, we shall be approaching as is depicted above in terms of the arrows. The depiction is of two parts. The first part is *Urdhva* (vertical) and the second part is *Tiryag* (crosswise). This as such, would help us comprehend as that geometric format of square is being divided into two triangles and only 1 out of two triangles is being utilised by the organisation format *of Sutra* 3.

In other words we require a minimum of three lines to have a close organisation to enclose planes (area) surface within its lines. The triangle as such is the printout of 3-space on 2-space. The sum of all the three internal angles of a triangle is p. The sum of all the three external angles of a triangle is 2p. We can express p as (3-2) p. The sum of all the four internal angles of a quadrilateral is 2p. We can express it as (4-2) p. In general, the printout of n-space on 2-space is n-polygon and the sum of the internal angles n-polygon is (n-2) p. This in general can be seen as that the printouts of n-space in the role of dimension on 2-space is n-polygon. Therefore, the organisation format *of Sutra* 3 is that of a printout of 3-space in the role of dimension on 2-space format. Like that, one can proceed to reach at the organisation formats of the entire range of the text of 16 *Ganita Sutras* and 13 *Upsutras* but the context and the space at hand would not permit to proceed ahead with it. However, the organisation formats achieved until this stage will indicate about the Mathematics of linear order of *Sutra* 1, spatial order of *Sutra 2* and of solid order *of Sutra* 3.

MANASARA

Manasara is a scripture that deals with space time frame. It covers a great deal on the science and art of sculpture and constructing buildings. Sri Prasana Kumar Acharya had reconstructed the text of *Manasara* by drawing the plates. He also prepared its translation in English.

The scripture begins with the prayer to Lord Brahma, the Creator and ends with the chiselling of third eye of the idol of Lord Shiva. The broad organisation of the scripture is of the range of 70 chapters with first eight chapters constituting a primary group-I.

The next ten chapters i.e., chapter 9 to 18 constitute a primary group-II. Then comes the central part of the *Manasara.* Chapter 19 to 30 cover single story building to 12 storied buildings. These 12 chapters constitute a central group-1. Next 20 chapters i.e., chapters 31 to chapter 50 cover central group-II and with it the subject of architecture as such is completed. Then follows the subject of sculptures.

The science of sculptures covered in chapters 51 to 70 can be organisationally divided in two sculptural groups. Chapters 51 to 65 constitute sculptural group-I and chapters 66 to 70 constitute sculptural group-II. The above topical division of the text has an organisational message of great importance as in terms of it we may reach at the geometric format of the organisation of the knowledge of the scripture. As the scripture begins with the prayer to Lord Brahma, the overlord of real 4-space and ends with the chiselling of third eye of Lord Shiva, the overlord of real 5-space, therefore, geometric format at the dimensional level is bound to be spatial with the flux of time being solid. In short, the space-time frame at the dimensional level is going to be E^2 (space) x E^3 (time). In terms of this dimensional order, we shall be manifesting working geometric domain within the space-time frame $(E^2)^4$ (space) x E^3 (solid time). Here, we shall be concentrating upon the concepts and comprehensions of dimensions of space and time in *Manasara.* In a way, we shall be taking up the topic of space, time and space-time in the light of Vedic wisdom. In the context, it may be relevant to note that the modern thought, Mathematics, science and technology is speculating the general space-time frame as E^3 (space) * E^1 (linear time). The modern thought, Mathematics, science and technology centre around linear dimensional reality but the Vedic systems avail multidimensional reality. The studies of the organisational formats of various Vedic scriptures reveal that higher dimensional geometric formats are being availed to organise the pure knowledge. Illustratively, we may take the case of the oldest book of mankind namely, *Sakla Rig Veda Samhita.*

RIG VED SAMHITA

Fortunately, *Rig Veda Samhita* is intact with us from first syllable to the last syllable and as the tradition goes, the whole range of Vedic knowledge is lively in this scripture of 432000 syllables, out of which 397265 syllables are manifest text while remaining 34735 syllables go deep as organisational format of the text and as such remain un-manifest. For the present, we may accept it an axiom that knowledge and organisation of knowledge are two distinct aspects of knowledge. Being scriptural text, we get the organised knowledge and as such both organisation format and the text are to be accepted as the knowledge content of the scripture. It is like a truck with goods yielding weight of the truck as well as of the goods loaded in the truck.

ORGANISATION OF RIG VEDA SAMHITA

The total knowledge contents in *Rig Veda Samhita* are 432000 syllables and the manifest text is of 397265 syllables. The *Rig veda Samhita* consists of:

1. *Mandals* 10
2. *Ashtaks* 08
3. Chapters 64
4. *Anuvaks* 85
5. *Suktas* 1028
6. *Vargas* 2024
7. *Richas* 10552

Mathematical basis of the organisational format of the *Rig Veda Samhita* reveals that the Vedic knowledge is organised on geometric format of real 6-space. It admits 4-space in the role of dimension while modern thought, Mathematics, science, and technology centre around 3-space reality and as such *Vedas* are invincible fort for the modern mind. As such, we have to learn and understand the Vedic wisdom. For this we have to re-examine the rationale and basis of our axioms. It is only by approaching the Vedic knowledge, the Vedic way, that we may have real bliss of Vedic wisdom.

3

Addition

Addition is the basic operation in Mathematics.

There are several methods for adding numbers in Vedic Mathematics for addition.

Method 1: Adding from left to right

In conventional method, teachers are teaching addition from right to left. In Vedic Mathematics we can do addition from left to right which is more quicker than conventional method.

Example 1.

```
  23
+ 41
----
  64
```

In conventional method, first we add 3+1(=4), and then add 2+4 (=6).

Example 2.

```
   8  9
+  2  8
-------
  10 17
```

- →In Vedic Mathematics, first we add the left side digits, 8 and 2(=10) and add 9 +8 (=17). Take 1 as carry from 17 and add to 10.

i.e., Answer is 117

Example 3.

```
   1  4  7
+  3  2  8
----------
   4  6 15
```

Step 1. 8 + 7 = 15

Step 2. 4 + 2 = 6

Step 3. 3 + 1 = 4

Step 4. Take leading 1 of 15 as carry and add the same to 6. ie. 6+1(=7)

Answer is 475.

Example 4.

1	4	6	7
+8	7	9	3
9	11	15	10

Step 1. Add each digits from left to right as shown in above figure.
Step 2. Start from right, take left digit of each number as carry (if it is a single digit number, there will be no carry) and add carry of each number to the previous digits.

Take 1 as carry from 10 and add 1 to 15(=16), take 1 from 16 as carry and add to 11(=12), take 1 again as carry from 12 and add to 9(10)

Answer is 10260

Example 5. Add 489653 and 658974

4	8	9	6	5	3
+6	5	8	9	7	4
10	13	17	15	12	7

Step 1. Add each digits from left to right as shown in above figure.
Step 2. Start from right, take left digit of each number as carry (if it is a single digit number, there will be no carry and add carry of each number to the previous digits.

Answer is 1148627

Example 6. Add 8965321 and 4789651

8	9	6	5	3	2	1
+4	7	8	9	6	5	1
12	16	14	14	9	7	2

Step 1. Add each digits from left to right as shown in above figure.

Step 2. Start from right, take left digit of each number as carry (if it is a single digit number, there will be no carry) and add carry of each number to the previous digits.

Answer is 13754972

Try these.

(1) 45 + 69 (2) 74 + 12

(3) 12 + 36 (4) 985 + 42

(5) 541 + 235 (6) 412 + 812

(7) 325 + 347 (8) 541 + 635

(9) 258 + 369 (10) 457 + 698

Answers:

1) 114 2) 86 3) 48 4) 1027 5) 776 6) 1224 7) 672 8) 1176 9) 627 10) 1155

Method 2 : Ekadhikena purvena method

This is another simple method for adding numbers used in Vedic Mathematics.

Step 1. Add the digits from left in column wise.

Step 2. If the sum is greater than or equal to 10, put a dot to above the number

Step 3. Count the number of ticks and place it down to the number next to the unit place and sum the two.

Example 1. 45 + 89

```
    4  5
    .  .
    8  9
  -------
    2  4
 1  1
  -------
 1  3  4
```

Step 1. Add 9 and 5 (=14). Put a dot above 9 and write only 4

Step 2. Add 8 and 4 (=12). Put a dot above 8 and write only 2

Step 3. Count the number of dots in each digits and write down the count from next to unit place of the answer. Here we have 1 dots above each digits, so write down two 1s in the tenth place and hundredth place in figure.

Step 4: Add the digits from right in column wise.

Answer is 134

Example 2. 478 + 698 + 354

```
    4  7  8
    .  .  .
    6  9  8
       .  .
    3  5  4
   ---------
    3  1  0
 1  2  2
 -----------
 1  5  3  0
```

Step 1. Add 8 and 8(=16), put a dot above 8 and add the unit digit 6 of 16 with 4(=10). Put a dot above 4 and write 0 as the first digit of answer.

Step 2. Add 7 and 9(=16), put a dot above 9 and add the unit digit 6 of 16 with 5(=11). Put a dot above 5 and write 1 in the answer column.

Step 3. Add 4 and 6(=10), put a dot above 6 and add the unit digit 0 of 10 with 3(=3). Write 3 in the answer column.

Step 4. In the first column, we have 2 dots, write 2 in the second position(below 1) of answer row.

Step 5. In the second column also, we have 2 dots, write 2 in the third position(below 3).

Step 6. In the third column, we have only one dot, so write 1 in the 4th position. Add the digits from right in column wise.

Answer is 1530.

Example 3. 6548 + 4587 + 6987 + 3578

	6	5	4	8
	.	.	.	.
	4	5	8	7
			.	.
	6	9	8	7
		.		.
	3	5	7	8
	9	4	7	0
1	2	2	3	
1	11	6	10	0
2	1	7	0	0

Answer of 6548 + 4587 + 6987 + 3578 is 21700

Try these:

(1) 125 + 368

(2) 456 + 698

(3) 4578 + 3654

(4) 3651 + 9531

(5) 21456 + 11569

(6) 21456 + 98521

(7) 1236547 + 369851 + 698541

(8) 1245678 + 3215478 + 6214783 + 3214563

(9) 12365412 + 9853641

(10) 145698 + 3256 + 698745 + 21456 + 6981

Answers :

1) 493 2) 1154 3) 8232 4) 13182 5)33025

6)119977 7) 2304939 8) 13890502 9) 22219053

10) 876136

4

Subtraction

Subtraction is the mathematical operation that represents removing objects from a collection.

Number of sutras are available in Vedic Mathematics for subtraction.

Subtraction from a power of 10.

Method : All from 9 and the last from 10

This is the method of subtracting all the numbers except last digit from 9 and last digit has to be subtracted from 10.

* If the number of digits of subtrahend and number of zeros in minuend are not equal, put zeros in the left to subtrahend to make the digits equal.

* If the subtrahend has any zeros from right, then write zeros in the answer as well.

Applying the formula 'All from 9 and the last from 10' to any number especially the big one's reduces it to its smaller counterpart that can be easily used for calculations involving the big digits like 7, 8, and 9.

Example 1 : Subtract 32 from 100.

Step 1. In this example, minuend is 100 and subtrahend is 32. Number of zeros in minuend and number of digits of subtrahend are equal.

Step 2. Subtract all the digits subtrahend from 9 except first digit from right(i.e., unit place digit). Subtract 1st digit(unit place digit) from 10. i.e., subtract 3 from 9(=6) and 2 from 10(8).

Answer is 68.

Example 2 : Subtract 574 from 1000

Step 1. Subtract all the digits from 9 except first digit from right(i.e., unit place digit). Subtract 1st digit from 10.

i.e., subtract 5 from 9(=4), 7 from 9(=2) and 4 from 10(=6).

Answer is 426.

Example 3 : Subtract 54 from 1000.

Step 1. Since 1000 has 3 zeros and 54 has only 2 digits. Write 54 as 054. Then subtract 0 from 9(=9), 5 from 9(=4) and 4 from 10(=6). i.e., 946

Answer is 946

Example 4 : Subtract 16 from 100000.

Step 1. Since 100000 has 5 zeros, and 16 has only 2 digits, write 16 as 00016. Subtract 0 from 9(=9), 0 from 9 (=9), 0 from 9(=9), 1 from 9 and 6 from 10.

i.e., 99984.

Answer is 99984.

Example 5 : Subtract 9854 from 10000

Step 1. Here number of zeros of minuend and number of digits of subtrahend are equal. So no need to add zeros in subtrahend. Subtract 9 from 9(=0), 8 from 9 (=1), 5 from 9 (=4) and 4 from 10 (=6). i.e., 0146

Answer is 146.

Example 6 : Subtract 35689 from 100000

Step 1 : Subtract 3 from 9, 5 from 9, 6 from 9, 8 from 9 and 9 from 10

i.e., 64311

Answer is 64311

Example 7 : Subtract 450 from 1000.

Step 1. Since the first digit from right (unit place) is a zero, put that zero as it is in answer also.

Here first digit from right is zero, put zero as it is in the answer. Then subtract 4 from 9 and 5 from 10.

Answer is 550.

Try these.

(1) 100 – 65	(2) 1000 – 654
(3) 1000 – 984	(4) 10000 – 9841
(5) 1000 – 45	(6) 100 – 8
(7) 100000 – 32	(8) 100000 – 654
(9) 10000 – 540	(10) 1000 – 310

Answers:

1) 35 2) 346 3) 16 4) 159 5) 955 6) 92
7) 99968 8) 99346 9) 9460 10) 690

Subtracting from a Multiple of a Base

Method : All from 9 and last from 10, and one less than the one before.

Example 1 : Subtract 16 from 700.

In this example, 700 is Minuend and 16 is the Subtrahend.

Step 1. Check the number of minuend and subtrahend. If the number of digits are not equal, put zeros to left of Subtrahend. Then 16 becomes 016.

Step 2. Take the first digit from left of minuend, subtract 1 from that number. Here, 1 is subtracted from 1(=6). 6 is the first digit of the answer from left.

Step 2. Take as much digits as number of zeros in the minuend. Then follow the method 'all from 9 and last from 10'. Here minuend has 2 zeros, take 16 from subtrahend and subtract 1 from 9, 6 from 10. i.e., 84

Step 4. Combine both the answers got by step 2 and step 3, (6 and 84), we get the answer.

Answer is 684.

Example 2 : Subtract 125 from 800

Here number of digits of minuend and subtrahend are equal. In this case where both numbers have the same number of digits reduce the first digit of the first number(minuend) by one more than the first digit of the second number(subtrahend) to get the first digit of the answer.

Step 1. Count the number of digits of minuend and subtrahend. Here both are equal.

Step 2. First digit of the minuend is 8. Subtract one more than the first digit of subtrahend (here first digit is 1). i.e., Subtract 2 from 8 (=6).

Step 3. 800 has only 2 zeros, take leading 25 of 125, do the calculations using the method 'all from 9 and last from 10', subtract 2 from 9 and 5 from 10. i.e., 75

Step 4. Combine both the answers of step 1 and step 2, answer becomes 675.

Answer is 675.

Example 3 : subtract 654 from 6000

Step 1. Since minuend has 4 digits and subtrahend has only 3 digits, write 654 as 0654.

Step 2. Subtract 1 from 6(=5).

Step 3. Since 1000 has 3 zeros, using the method 'all from 9 and last from 10', subtract 6 from 9, 5 from 9 and 4 from 10. We get 346.

Combining both answers of step 2 and step 3, answer becomes 5346.

Answer is 5346.

Example 3 : Subtract 453 from 70000

Step 1. To get first part of the answer, less 1 from 7 (i.e., 7-1 = 6).

Step 2. 70000 has 4 zeros and 453 has only 3 digits, append 0 to the first of 453. i.e., 0453. Using the method 'all from 9 and last from 10', subtract 0 from 9, 4 from 9, 5 from 9 and 3 from 10. We get, 9547.

Step 3. Combining both the answers of step 1 & 2, we get 69547.

Answer is 69547.

Example 4 : Subtract 5412 from 15000

Step 1. To get the first part of the answer, first add 1 to 5 making to 6, and less 6 from 15, making the first part of the answer to 9.

Step 2. Since 15000 has 3 zeros, take 3 digits from unit place, i.e., using the method 'all from 9 and last from 10', subtracting 4 from 9, 1 from 9 and 2 from 10, we get 588.

Answer is 9588.

Example 5 : Subtract 3214 from 180000

Step 1. To get the first part of the answer, reduce 18 by 1 which is equal to 17.

Step 2. As per the method of 'all from 9 and last from 10', subtracting 3 from 9, 2 from 9, 1 from 9 and 4 from 10, we get 6786

Answer is 176786.

Try these:

(1) 700 – 254
(2) 1500 – 478
(3) 8000 – 654
(4) 400 – 15
(5) 9000 – 4789
(6) 30000 – 654
(7) 200000 – 65412
(8) 12000 – 456
(9) 8000 - 569
(10)6000 - 698

Answers:

1) 446 2) 1022 3) 7346 4) 385 5) 4211 6)29346
7) 134588 8)11544 9) 7431 10) 5302

Subtracting Near a base

When subtracting a number close to multiples of 10, first subtract from multiples of 10 and do the correction on answer accordingly.

Example 1 : Subtract 19 from 55

19 is close to 20, just 1 less than 20, subtract 20 from 55 (=35), then add 1 to 35, i.e., 36.

55 – 19 = (55 – 20) + 1 = 36

Answer is 36.

Example 2 : Subtract 38 from 67

38 is close to 40, just 2 less than 40, subtract 40 from 67 (=27), then add 2 to 40.

67 - 38 = (67- 40)+ 2 → 27 +2 = 29.

Answer is 29.

Example 3 : Subtract 168 from 197

168 is close to 170, just 2 less than 170, subtract 170 from 197 (=27), then add 2 to 27.

197 - 168 = (197 - 170) +2 =27 + 2 =29

Thus answer is 29.

Try these:

(1) 65 – 39	(2) 80 – 36
(3) 90 – 45	(4) 490 – 28
(5) 1900 – 859	(6) 19000– 119
(7) 22990 – 1598	(8) 19960 – 9856
(9) 29990 – 5489	(10) 98900 – 9865

Answers :

1) 26 2) 44 3) 45 4) 462 5) 1041 6)18881

7) 21392 8)10104 9) 24501 10) 89035

Subtraction using complements.

Example 1 : Subtract 19 from 37

```
3 7
1 9
---
1 8
```

Step 1. 7 – 9 is not possible with (+) sign, so take the complement of 9 and add it with 7. To take complement, subtract the digit from 10(here 10 - 9 = 1) i.e., 1 + 7 = 8. 8 is the first digit of the answer from right.

Step 2. While taking the complement, add 1 to the previous digit. i.e., 1 + 1 = 2. Then subtract 2 from 3. ie., 3 – 2 = 1.

Answer is 18.

Example 2 : 156 – 91

```
1 5 6
  9 1
-----
  6 5
```

Step 1. 6 – 1 = 5

Step 2. 5 – 9 is not possible, so take the complement of 9 and add it to 5. Complement of 9 is 1. Add 1 to 5(=6)

Step 3. Since we have taken complement, add 1 to the digit left to 9. Here it is zero. 0 + 1 = 1. Then 1 – 1 = 0.

Answer is 65.

Example 3. 842 – 125

```
8 4 2
1 3 5
-----
```

Step 1. Since 2 – 5 is not possible, so take the complement of 5 and add to 2. Complement of 5 is 5(10-5=5) and add 5 to 2 (= 7).

Step 2. 4-3 = 1

Step 3. 8-1 =7

Thus, 842 – 125 =717.

Try these:

(1) 756 – 635

(2) 985 – 638

(3) 9865 – 6352

(4) 6985 – 3687

(5) 10589 – 6547

(6) 29641 – 23658

(7) 10589 – 6359

(8) 7415 – 6985

(9) 21458 – 16985

(10) 15985 – 6398

Answers:

1) 121 2) 347 3) 3513 4) 3298 5)4042 6) 5983

7) 4230 8)430 9) 4473 10) 9587

5

Multiplication

Ekanyunena Purvena of Vedic Mathematics is applicable whenever the multiplier completely consists of nines.

Either Multiplicand or multiplier should be entirely consisting of 9's. But for simplicity we will call number with all 9's as multiplier.

Ekanyunena Purvena Sutra Multiplication has below 3 cases:

I. Case 1 :

Digits of multiplicand and multiplier are same.

- **1st step.** 1 less of multiplicand.
- **2nd step.** Complement of multiplicand or (Multiplier – previous column).

Examples:

1. 8×9

Here 8 is multiplicand and 9 is multiplier.

Step 1. 1 less of multiplicand. i.e., 8 – 1 = 7. (7 is the first digit from left of the answer.)

Step 2. Complement of multiplicand. Complement of 8. i.e., 10 – 8 = 2.(2 is the second digit from left of the answer)

8 x 9 =72 is the answer.

2. 76×99

Step 1. 1 less of multiplicand. i.e., 76 – 1 = 75(75 is the first two digits from left of the answer)

Step 2. Complement of multiplicand. Complement of 76. i.e., 100 – 76 = 24. 24 is the last digits from left of the answer.

76 × 99 =7524 is the answer.

3. 146 × 999

Step 1. 146 – 1 = 145

Step 2. Complement of 146 i.e., 1000 – 146 = 854

Answer is 145854.

4. 2313 × 9999

Step 1. 2313 – 1 = 2312.

Step 2. Complement of 2313. i.e.,10000 – 2313 = 7687

2313 × 9999 = 23127687 is the answer.

5. 42315 × 99999

Step 1. 42315 – 1 = 42314

Step 2. Complement of 42315 i.e., 100000 – 42315 = 57685

Thus answer is 4231457685

Try these:

(1) 48 × 99	(2) 159 × 999
(3) 36 × 99	(4) 789 × 999
(5) 1458 × 9999	(6) 6985 × 9999
(7) 12365 × 99999	(8) 98536 × 99999
(9) 12365 × 99999	(10) 47896 × 99999

Answers:

1)4752 2)158841 3) 3564 4) 788211 5) 14578542

6) 69843015 7) 1236487635 8) 9853501464

9)1236487635 10) 4789552104

II: Case 2:

Multiplier (number with 9's) has digits less than multiplicand.

Count the digits of multiplier, if it is 1

- **Step 1.** Count the digits of multiplier. Split the multiplicand as take as much digit from right as one group and rest of the number as second group.

- **Step 2.** Add 1 to the number of second group.
- **Step 3.** Subtract the answer of step 2 from multiplicand.
- **Step 4.** Take compliment of first group of multiplicand.

Examples:

1. 14 × 9

Step 1. Number of digits of multiplier is 1. Split multiplicand as 1(second group) and 4(first group).

Step 2. Add 1 to second group(1+ 1=2).

Step 3. Subtract 2 from 14(=12).

Step 4. Complement of first group digit is 6.

14 x 9 = 126 is the answer.

2. 126 × 9

Step 1. Number of digits of multiplier is 1. Split multiplicand as 12(second group) and 6(first group).

Step 2. Add 1 to 12(=13).

Step 3. Subtract 13 from 126(=113)

Step 4. Complement of 6 is 4.

126 x 9 = 1134 is the answer.

3. 1689 × 99

Step 1. Number of digits of multiplier is 2. Split multiplicand as 16(second group) and 89(first group).

Step 2. Add 1 to 16(=17).

Step 3. Subtracts 17 from 1689 i.e., 1672.

Step 4. Complement of 89 i.e., 11

1689 x 99 = 167211 is the answer.

4. 51234 × 99

Step 1. Number of digits of multiplier is 2. Split multiplicand as 512(second group) and 34 (first group).

Step 2. Add 1 to 512(=513)

Step 3. Subtract 513 from 51234 i.e., 50721

Step 4. Complement of 34. i.e., 100 – 34 = 66

51234 × 99 = 5072166 is the answer.

5. 675643 × 999

Step 1. Since multiplier has three digits, split the multiplicand as 675 (second group) and 643 (first group).

Step 2. Add 1 to 675(=676).

Step 3. Subtracts 676 from 675643 i.e., 674967

Step 4. Complement of 643. i.e., 1000 – 643 = 357

675643 × 999 = 674967357 is the answer.

6. 1234869 × 9999

Step 1. Since multiplier has four digits, split the multiplicand as 123 (second group) and 4869 (first group).

Step 2. Add 1 to 123(=124).

Step 3. Subtracts 124 from 1234869 i.e., 1234745

Step 4. Complement of 4869 i.e., 10000 – 4869 = 5131

1234869 × 9999=12347455131 is the answer.

Try these:

(1) 1115 × 99
(2) 459 × 99
(3) 4569 × 99
(4) 6985 × 99
(5) 4789 × 999
(6) 365478 × 999
(7) 45698 × 999
(8) 12365 × 9999
(9) 74896 × 999
(10) 12365478 × 99999

Answers :

1) 110385 2) 45441 3) 452331 4) 691515 5) 4784211

6) 365112522 7) 45652302 8) 123637635

9) 74821104 10) 1236535434522

III: Case 3 :

Multiplier (number with 9's) has digits more than multiplicand.

- **1st step.** 1 less than Multiplicand
- **2nd step.** Subtracts multiplicand from multiplier +1

Examples:

1. 4×99

   Step 1. 1 less than multiplicand i.e., 3

   Step 2. 100 – 4 = 96 (subtract multiplicand from 99 + 1)

   Thus answer is 396.

2. 23×999

   Step 1. 1 less than multiplicand i.e., 22

   Step 2. 1000 – 23 = 977(subtract multiplicand from 999 + 1)

   Thus answer is 22977

3. 456×99999

   Step 1. 1 less than multiplicand i.e., 455

   Step 2. 100000 – 456 (Subtract multiplicand from 99999 + 1)

   Thus answer is 45599544

4. 4567×999999

   Step 1. 1 less than the multiplicand i.e., 4566

   Step 2. 1000000 – 4567 = 995433(Subtract multiplicand from 999999 + 1)

   Thus answer is 4566995433

5. 858 x 9999

   Step 1. 1 less than the multiplicand i.e., 857

Step 2. 10000 – 858 = 9142 (Subtract multiplicand from 999999 + 1)

Thus answer is 8579142.

Try these :

(1) 12 × 999 (2) 56 × 9999

(3) 896 × 99999 (4) 7458 × 99999

(5) 745 × 99999 (6) 129 × 99999

(7) 4598 × 99999 (8) 47896 × 999999

(9) 9655 × 9999999 (10) 12569 × 999999

Answers :

1) 11988 2) 559944 3) 89599104 4) 745792542

5) 74499255 6) 12899871 7) 459795402

8) 47895952104 9) 96549990345 10) 12568987431

Tirthaji Maharaj has classified tricks to Multiply Numbers in Vedic Mathematics in Specific and General Methods. Specific Multiplication Methods can be applied when numbers satisfy certain conditions like both numbers closer to 100 or numbers closer to each other or addition of last digits of both numbers is 10, etc. While General Multiplication Methods can be applied to any types of numbers.

Depending on Specific and General Techniques, Multiplication in Vedic Mathematics are classified in the form of Sutras as below. Let us see the Vedic Mathematics Multiplication techniques.

1. Nikhilam Sutra (Specific Technique)
2. Anurupyena Sutra (Specific Technique)
3. Urdhva Tiryak Sutra and Vinculum Process (General Technique)
4. Ekayunena Purvena (Specific Technique)
5. Antyaordaske'pi (Specific Technique)

Nikhilam Sutra:

This is the simplest trick to multiply numbers using Vedic Mathematics.

Using Nikhilam Sutra it is simpler to multiply numbers like 98 & 95, 997 & 987, 102 & 112, 995 & 1008 i.e. the numbers which are closer to power of 10. This Sutra is a Specific method of Multiplication in Vedic Mathematics which shows shortcuts to multiply numbers which are closer to power of 10 (10, 100, 1000, etc.)

This will generate 3 cases:

CASE 1 : Numbers closer and lesser than power of 10.

Examples

1. Multiply 7 by 6

   Here base is 10.

   Step 1.

7 – 3	7 is 3 less than 10
6 – 4	6 is 4 less than 10

   Step 2.

   Multiply vertically i.e., –3 *– 4 = 12. 12 is a 2 digit number, so take leading 1 of 12 as a carry. 2 is the first digit of the answer from right.

   Step 3.

   Cross subtract the digits 7 and 4 or 6 and 3. 7 – 4 = 3 and 6 – 3 = 3, both answers are same. Add carry 1 of the step 1 to 3(=4). 4 is the second digit of the answer from right.

7	–3
6	–4
3	(1) 2

   $6 \times 7 = 42$.

2. Multiply 97 by 94 Here base is 100.

Step 1.

97 – 3 97 is 3 less than 100

94 – 6 94 is 6 less than 100

Step 2.

Multiply vertically – 3 * – 6 = 18. 18 is 2 digit number and base is also 100.

So here is no carry to forward. 18 is the first figure of the answer from right.

Step 3.

Cross subtract 97 and 6 or 94 and 4 to get the second figure of the answer from right.

97 – 6 = 91 or 94 – 3 = 91 both are same. 91 is the second figure of the answer.

97	–3
94	–6
91	+18

Answer of 97 × 94 is 9118

3. Multiply 91 by 99. Here base is 100.

Step 1.

91 – 9 91 is 9 less than 100

99 – 1 99 is 1 less than 100

Step 2.

Multiply vertically – 9 * – 1 = 9. Since our base is 100, and 9 is a single digit number, so needs to add 0 to make 9 as a 2 digit number. First figure of the answer is 09.

Step 3.

Cross subtract to get the second figure of the answer.

91 – 1 = 90 and 99 – 9 = 90 both answers are same.

Step 3.

91 −9

99 −1

90 09

Answer of 91 × 99 is 9009.

4. Multiply 989 by 986 Base is 1000

Step 1.

989 – 11 989 is 11 less than 1000

986 – 14 986 is 14 less than 1000

Step 2.

Multiply vertically – 11 * – 14 = 154. Base is 1000 and 154 is also have 3 digits, so no carry. 154 is the first figure of the answer.

Step 3.

Cross subtract to get the second figure of the answer.

989 – 14 = 975, 986 – 11 = 975 both are same. 975 is the second figure of the answer.

989 −11

986 −14

975 154

Answer of 989 × 986 is 975154.

5. 994 × 988 base is 1000

Step 1.

994 – 6 994 is 6 less than 1000

988 – 12 988 is 12 less than 1000

Step 2.

Multiply vertically – 6 * – 12 = 72. Base is 1000 but 72 has only 2 digits, add 0 before 72 to get the first figure.

Step 3.

Cross subtract to get the first figure.

994 – 12 = 982, 988 – 6 = 982 both are same. 982 is the second figure of the answer.

994	–6
988	–12
982	0 72

Answer of 994 × 988 is 982072.

6. 775 × 985 base is 1000

Step 1.

775 – 225 775 is 225 less than 1000

985 – 15 985 is 15 less than 1000

Step 2.

Multiply vertically to get the first figure.

– 225 * – 15 = 3375, Base is 1000, so take only 3 digits of 3375, trailing digit of 3375 is 3. Take 3 as carry to next step.

Step 3.

Cross subtract to get the second figure of the answer.

775 – 15 = 760, 985 – 225 = 760, both are same. Add carry 3 of step 2 to 760 (=763). 763 is the second digit of the answer.

775	–225
985	–15
760	3 375

Answer of 775 × 985 is 763375.

Try these :

(1) 5 × 6 (2) 93 × 98

(3) 91 × 99 (4) 985 × 991

(5) 996 × 992 (6) 991 × 993

(7) 9991 × 9996 (8) 9999 × 9994

(9) 996 × 992 (10) 999995 × 999998

Answers :

1) 30 2) 9114 3) 9009 4) 976135 5) 988032

6) 984063 7) 99870036 8) 99930006 9) 988032

10) 999993000010

CASE 2 : Both the numbers are greater than power of 10.

The method and rules follow as they are. The only difference is the positive deviation instead of cross subtract we follow cross add.

1. 14 × 17 Base is 10

Step 1.

14 + 4 14 is 4 more than 10

17 + 7 17 is 7 more than 10

Step 2.

Multiply vertically, 4 × 7 = 28. Here base is 10. Trailing digit of 28 is 2. Take 2 as a carry and add 2 to the answer of next step. 8 is the first figure of answer from right.

Step 3.

Cross add to get the second figure.

14 + 7 = 21, 17 + 4 = 21 both are same.

Add carry 2 of step 2 to 21(=23). 23 is the second figure of the answer.

14	+4
17	+7
21	2 8

Answer of 14 × 17 is 238.

2. 103 × 108 base is 100

Step 1.

103 + 3   103 is 3 more than 100.

108 + 8   108 is 8 more than 100.

Step 2.

Multiply +3 by +8. i.e., 24. Base is 100 and number of digits in 24 is also 2. So no carry to take forward. 24 is the first digit of the answer.

Step 3.

Cross add to get the second figure of the answer.

103 + 8 = 111 or 108 + 3 = 111

103	+3
108	+8
111	24

Answer of 103 × 108 = 11124.

3. Multiply 1009 by 1012.

Base is 1000.

Step 1.

1009 + 9   1009 is 9 more than 1000.

1012 + 12   1012 is 12 more than 1000.

Step 2.

Multiply vertically + 9 by + 12. i.e., 108. Base is 1000 and 108 has 3 digits. So no carry. 108 is the first figure of the answer.

Step 3.

Cross add to get the first figure.

1009 + 12 = 1021 or 1012 + 9 = 1021. Both are same. 1021 is the second figure of the answer.

1009	+9
1012	+12
1021	108

Answer of 1009 × 1012 = 1021108.

4. Multiply 10015 × 10009

Step 1.

10015 + 15   10015 is 15 more than 10000

10009 + 9   10009 is 9 more than 10000

Step 2.

Multiply vertically + 15 * +9 = 135. Base is 10000 and 135 has only 3 digit, so add 0 in trailing of 135. Thus 135 becomes 0135 and 0135 is the first figure of the answer from right.

Step 3.

Cross add to get the first figure.

10015 + 9 = 10024 or 10009 + 15 = 10024

10015	+15
10009	+9
10024	0135 → 100240135

Answer of 10015 × 10009 is 100240135.

5. Multiply 11120 × 11155

Step 1.

11120 + 1120   11120 is 1120 more than 10000

11155 + 1155   11155 is 1155 more than 11155

Step 2.

Vertically multiply 1120 * 1155 = 1293600. Base is 10000 so need only 4 digits in the first part of the answer. Take trailing 129 as carry and add to the answer of next step.

Step 3.

Cross add 11120 + 1155 = 12275 or 11155 + 1120 = 12275.

Both are same. Add carry of step 2 to 12275.

i.e., 129 + 12275 = 12404

11120	+1120
11155	+1155
12275	(129) 3600

Answer of 11120 × 11155 is 124043600.

Try these :

(1) 13 × 15 (2) 106 × 110

(3) 115 × 120 (4) 1005 × 1003

(5) 1020 × 1025 (6) 10006 × 10012

(7) 1010 × 1005 (8) 1006 × 1009

(9) 10008 × 10015 (10) 10025 × 10010

Answers :

1) 195 2) 11660 3) 13800 4) 1008015 5) 1045500

6) 100180072 7) 1015050 8) 1015054 9) 100230120

10) 100350250

CASE 3 : One number is above and the other below the base.

In this case, one deviation is positive and the other is negative. So the product of deviations becomes negative and the right hand side of the product of deviations becomes negative and the right hand side of the answer obtained will therefore have to be subtracted.

Example

1. 12 × 8 Base is 10

Step 1.

12 + 2 12 is 2 more than base 10.

8 – 2 8 is – 2 less than 10.

Step 2.

Multiply vertically +2 * –2 = –4.

We must convert this negative number by replacing it

with it's compliment. Complement of 4 is 6(i.e., 10 – 4 = 6). So replace –4 with 6. Whenever we do a replacement of this type, we must also subtract one from the answer of next step. First figure of the answer is 6.

Step 3.

Cross add taking care of sign of digits. i.e., 12 + –2 = 10.

8 + +2 =10 both are same. As mentioned in step 1, subtract 1 from 10(=9). Second figure of answer is 9.

12	+2
8	–2
10	–4

9 6 (complement of 4 is 6, subtract 1 from 10 as we have taken complement of 4.)

Answer of 12 × 8 = 96.

2. 14 × 7 Base is 10

Step 1.

14 + 4 14 is 4 more than base 10

7 – 3 7 is 3 less than base 10

Step 2.

Multiply vertically +4 * –3 = –12.

We must convert this negative number by replacing it with it's compliment. In this case base is 10, so take the complement of first digit only. The other digit is carry forwarded. In this case, first digit of –12 is 2. Complement of 2 is 8. The other digit -1 is carry forwarded. Whenever we take complement, we have to subtract 1 from the next part. Thus, in this case we have to subtract 2 from the next part of the answer. 8 is the first part of the answer.

Step 3.

Cross add taking care of sign of the digits.

12 + –3 = 11 or 7 + +4 = 11 both are same.

As mentioned in step 2, subtract 2 from 11. i.e., 11 – 2 = 9. is the second part of the answer and 8 is the first part.

14	+4
7	–3
11	–12

9 8

(complement of first digit is 8 and -2 is carry forwarded (-1 for complement and the other for leading digit of -12). 11 – 2 = 9

Answer of 14 × 7 = 98.

3. 110 × 95 Base is 100

Step 1.

110 + 10 110 is 10 more than 100

95 - 5 95 is 5 less than 100

Step 2.

Multiply vertically +10 and -5 (= -50). Take the complement of 50 (i.e., 100 - 50 = 50). Since we have taken complement, take -1 as carry to the answer of next step. 50 is the first part of the answer.

Step 3.

Cross add taking care of sign of the digits.

110 + -5 = 105 or 95 + 10 =105 both are same. As mentioned in step 2, add carry -1 to 105(105 + -1 = 104). 104 is the second part of the answer.

```
110   +10
 95    −5
----------
105   -50
```

(Completment of 50 is 50. Add carry -1 to 105(=104))

Answer of 110 * 95 is 10450

4. 997 × 1005 Base is 1000

Step 1.

997 -3 997 is 3 less than 1000

1005 +5 1005 is 5 more than 1000

Step 2.

Multiply vertically -3 and +5 (=-15). Complement of 15 is 985 (i.e., 1000-15 = 985). Since we have taken complement, take -1 as carry to answer of next step. 985 is the first part of the answer.

Step 3. Cross add taking care of sign of the digits.

(997 + 5) or (1005 + – 3) = 1002. Add carry -1 of step 2 1002(=1001). 1001 is the second part of the answer.

Answer of 997 × 1005 is 1001985.

Try these :

1) 10 * 8 2) 98 * 108 3) 992 * 1010 4) 9998 * 10005
5) 9990 * 10010

Answers:

1) 80 2) 10584 3) 1001920 4) 100029990 5) 99999900

Urdhva Tiryak Sutra:

This is another great shortcut method of multiplication using Vedic Mathematics. Urdhva Tiryak is General method of multiplication in Vedic Maths which provides shortcut to multiply any types of numbers.

(a) Multiplication of 2 digit numbers

Example 1. 14 × 12

Step 1. Multiply right hand most digit of multiplicand with right hand most digit of multiplier.

In this case, multiply 4 and 2 i.e., 8. This is the first digit of the answer.

Step 2. Diagonally multiply first digit of multiplicand with second digit of multiplier. Also, diagonally multiply second digit of multiplicand with first digit of multiplier. Add these two answers.

In this case, multiply 4 x 1, i.e., 4 and multiply 1 × 2 i.e., 2, add 4 and 2, i.e., 6

Second digit of the answer is 6.

Step 3. Multiply second digit of multiplicand with second digit of multiplier.

Here 1 × 1= 1. This is the third digit of the answer.

So answer is 168.

Symbolically,

1 4 ×

1 2

1 × 1 (1 × 2 + 4 × 1) 4 × 2 -> 168

So, 14 × 12 = 168.

Example 2. 23 × 34

2 3

3 4

Step 1. Multiply 3 × 4 -> 12, Carry forward leading 1 of 12. First digit from right of the answer is 2.

Step 2. Multiply 2 × 4(=8) and 3 × 3(=9). Add these two values. i.e.,

8 + 9 = 17. Add the carry forwarded 1 of step 1 to 17. i.e.18. Carry forward leading 1 of 18. Second digit from right is 8.

Step 3. 2 × 3 = 6. Add the carry forwarded 1 to 6 i.e., 7.

Arranging the digits, answer of 23 x 34 is 782.

Example 3. 46 × 87

Step 1. Multiply 6 × 7, i.e., 42. Carry forward is 4 and first digit from right of answer is 2.

Step 2. Add sum of 4 × 7 and 8 x 6. i.e., 28 + 48 = 76. Add carry forwarded 4 of step 1 to 76. i.e., 80. 8 is carry forwarded, 0 is the second digit from right.

Step 3. 4 × 8 = 32, add carry forwarded 8 to 32. i.e., 40. Last digit from right is 40.

Arranging digits, answer of 46 x 87 is 4002.

Example 4. 47 × 35

Step 1. Multiply 7 and 5 i.e., 35. Digit 3 will be carry forwarded. First digit of answer from right is 5.

Step 2. Multiply 4 by 5 and 3 by 7. Add these two. i.e., 20 + 21 which is equal to 41. Digit 3 which is carry forwarded from step 1, is added to 41. i.e., 41 + 3 = 44. 4 is the second digit from right, 10th place digit 4 is carry forwarded.

Step 3. Multiply 4 by 3. i.e., 12. Number 4 which is carry forwarded from step 2 is added to 12. Then 12 becomes 16. 16 is the last digits from right.

So, 47 × 35 = 1645.

Try these :

(1) 48 × 56	(2) 78 × 96
(3) 71 × 54	(4) 49 × 95
(5) 69 × 91	(6) 96 × 63
(7) 28 × 93	(8) 99 × 96

(9) 41 × 94 (10) 85 × 96

Answers :

1) 2688 2) 7488 3) 3834 4) 4655 5) 6279 6) 6048

7) 2604 8) 9504 9) 3854 10) 8160

(b) Multiplication of 3 digit numbers

Example 1. 121 x 313

Step 1.

1 2 1
3 1 3

1 × 3 → 3 First digit of the answer from right is 3.
No carry.

Step 2.

1 2 1
3 1 3

2 × 3 + 1 × 1→ 7 Second digit of the answer from right is 7.

No carry.

Step 3.

1 2 1
3 1 3

1 × 3 + 1 × 3 + 2 × 1 → 8 Third digit from right is 8.

No carry.

Step 4.

1 2 1
3 1 3

1 × 1 + 2 × 3 → 7 4th digit from right is 7.

No carry.

Step 5.

1 2 1
3 1 3

$1 \times 3 \rightarrow 3$ 5th digit from right is 3.

Arranging the digits, answer of 121×313 is 37873.

Example 5. 446 x 587

Step 1.

4 4 **6**

5 8 **7**

$6 \times 7 \rightarrow 42$ First digit from right is 2 and carry is 4

Step 2.

4 **4** 6

5 8 **7**

$4 \times 7 + 8 \times 6$ + Carry from step $1 \rightarrow 28 + 48 + 4 \rightarrow 80$. 2nd digit from right is 0 and carry is 8

Step 3.

4 4 6

5 8 **7**

$4 \times 7 + 6 \times 5 + 4 \times 8$ + carry $\rightarrow 28 + 30 + 32 + 8 \rightarrow 98$ 3rd digit from right is 8, carry is 9.

Step 4.

4 4 6

5 **8** 7

$4 \times 8 + 4 \times 5$ + carry $\rightarrow 32 + 20 + 9 = 61$, 4th digit from right is 1, carry is 6.

Step 5.

4 4 6

5 8 7

4×5 + carry $\rightarrow 20 + 6 = 26$. Last digits from right is 26.

Arranging the digits, Answer is 261802.

Example 6 : 1234 x 2365.

Step 1.

1 2 3 **4**

2 3 6 **5**

$5 \times 4 = 20$

First digit from right place is 0 and 2 is carry forwarded.

Step 2.

1 2 **3** 4

2 3 6 **5**

$\mathbf{3 \times 5 + 6 \times 4 = 39}$

Second digit is 9 (+previous carry) and carry is 3. i.e., 9 + 2 = 11. Then 2nd digit changes to 1 and carry is changed to 3 + 1 =4

Step 3.

1 **2** **3** 4

2 3 **6** **5**

$2 \times 5 + 3 \times 4 + 3 \times 6 = 10 + 12 + 18 = 40$

3rd digit is 0 (+previous carry) and carry is 4

Then 3rd digit is 4 and carry is also 4.

Step 4.

1 2 **3** **4**

2 3 6 **5**

$1 \times 5 + 2 \times 4 + 2 \times 6 + 3 \times 3 = 5 + 8 + 12 + 9 = 34$

4th digit is 4(+previous carry) and carry is 3.

4th digit from right is 8 and carry is 3

Step 5.

1 2 **3** 4

2 3 6 5

$1 \times 6 + 2 \times 3 + 2 \times 3 = 6 + 6 + 6 = 18$

5th digit from right is 8(+previous carry) and carry in this step is 1. Then fifth digit is 8 + 3. i.e., 11. We cannot write a 2 digit number, so trailing 1 should also be add to carry.

So, 5th digit is changed to 1 and is changed to 2.

Step 6.

$$\begin{array}{cccc} \mathbf{1} & \underline{2} & 3 & 4 \\ \underline{\underline{2}} & 3 & 6 & 5 \\ \hline \end{array}$$

$1 \times 3 + 2 \times 2 = 7$

6th digit is 7(+previous carry), carry is 0 digit is changed to 9 and carry 0

Step 7.

$$\begin{array}{cccc} \mathbf{1} & 2 & 3 & 4 \\ \mathbf{2} & 3 & 6 & 5 \\ \hline \end{array}$$

$1 \times 2 = 2$

7th digit is 2(+previous carry). 7th digit is 2.

By arranging the values got from step 1 to step 7, we get 2918410.

Thus, $1234 \times 2365 = 2918410$.

Try these

(1) 12×45	(2) 125×657
(3) 987×321	(4) 235×3254
(5) 4587×3657	(6) 7894×3214
(7) 419×365	(8) 415×987
(9) 147×398	(10) 9521×1478

Answers :

1) 540 2) 82125 3) 316827 4) 764690 5) 16774659

6) 25371316 7) 152935 8) 409605 9) 58506

10) 14072038

Another method for Multiplication of 2-digit numbers by 2-digit numbers

The conventional method of multiplication involves the below steps.

$$\begin{array}{rr} a & b \\ \times x & y \\ \hline ay & by \\ ax & bx \\ \hline \end{array}$$

ax / (ay + bx) / by

Example1 : Replacing the letters with the numbers the above-mentioned example can be written as shown below:

(a) *(b)*

7 8

(x) *(y)*

3 2

ax / (ay + bx) / by

21/ 14 + 24 / 16

21/ 38 / 16

2496

The steps can be summarized as below.

Step 1. Start from the right-hand side, keep 6 (from 16) as the extreme right digit of the answer and carry forward 1.

Step 2. Then add the remainder 1 to the middle portion, 14 + 24 = 38. Then put 9 (from 39) as the answer digit and carry forward 3.

Step 3. Then add the remainder 3 to the digit on the extreme left, i.e. 21 to make it 24, which is comes out to be 2496.

Example 2.

$$\begin{array}{rr} 4 & 7 \\ \times 6 & 8 \\ \hline \end{array}$$

= 24/ 32 + 42/ 56

= 24/ 74/ 56

= 3196, 5 and 7 are carries / remainders at each stage.

Example 3.

$$\begin{array}{r} 7\ \ 8 \\ \times 2\ \ 8 \\ \hline \end{array}$$

= 14/56 + 16/64

= 14/72/64, 6 and 7 are the remainders at each stage.

= 2184

Try these.

(1) 45 × 56 (2) 78 × 96

(3) 73 × 69 (4) 19 × 92

(5) 36 × 78 (6) 69 × 87

(7) 69 × 73 (8) 73 × 55

(9) 86 × 96 (10) 96 × 78

Answers :

1) 2520 2) 7488 3) 5037 4) 1748 5) 2808

6) 6003 7) 5037 8) 4015 9) 8256 10) 7488

Multiplication of 3 digit number by 2 digit number

The conventional method for multiplication is as follows.

$$\begin{array}{r} 352 \\ \times 53 \\ \hline 1056 \\ 1760 \\ \hline 18656 \end{array}$$

It is further explained with alphabets.

a	b	c
×	x	y
ay	by	cy
ax	bx	cx

ax / ay + bx /vy + cx / cy

In the multiplication of 2-digits by 2-digits, there was only one cross, but here numbers of crosses are two.

Example 1 : 875 × 36

Let us write

a	b	c
8	7	5
	x	y
	3	6

ax/ay + bx/by + cx/cy

= 24/48 + 21/42 + 15/30

= 24/69 /57 /30   3, 6, 7 are the remainders at each stage.

=31500

Thus, 875 × 36 = 31500.

Example 2 : 459 × 96

4	5	9
	9	6

36 /24 + 45/30 + 81/54 -> by using ax/ay + bx/by + cx/cy

= 36/69/111/54

→ adding the carries to the previous digits, we get

459 × 96 = 44064

Answer of 459 × 96 is 44064.

Example 3 : 985 × 84

$$\begin{array}{r} 9\ 8\ 5 \\ 8\ 4 \\ \hline \end{array}$$

= 72 / 36 + 64 / 32 + 40/20 - > by using ax/ay + bx/by + cx/cy

= 72/100/72/20

→ Adding the carries to the previous digits, we get

= 82740

Thus, 985 × 84 = 82740.

Try these:

(1) 478 × 25 (2) 985 × 63

(3) 786 × 65 (4) 698 × 47

(5) 469 × 36 (6) 368 × 94

(7) 478 × 95 (8) 687 × 64

(9) 657 × 32 (10) 498 × 63

Answers :

1) 11950 2) 62055 3) 51090 4) 32806 5) 16884

6) 34592 7) 45410 8) 43968 9) 21024 10) 31374

Multiplication of 4-digit numbers by 2-digit numbers

Now you have learnt the multiplication of 2-digit numbers by other 2-digit numbers and three-digit numbers by two-digit numbers. Now you will learn the multiplication of 4-digit numbers by 2-digit numbers. The conventional method is,

$$\begin{array}{rrrrr} & a & b & c & d \\ & & & x & y \\ \hline & ay & by & cy & dy \\ ax & bx & cx & dx & \\ \hline \end{array}$$

ax / ay+bx /by+cx / cy +dx / dy

Example 1 :

```
      5    8    7    6
                4    5
    ------------------
     25   40   35   30
20   32   28   24
---------------------
```

= 20 / 25 + 32 / 40 + 28 / 35 + 24 / 30

=20 / 57 / 68 / 59 / 30

= 26 4 4 20 and 3, 6, 7, 6 are the remainders at each step.

It involves the following steps:

Step 1. Start from the right.

Step 2. Calculate $dy = 6 \times 5 = 30$, keep 0 as the right-hand digit of the answer and carry forward 3.

Step 3. Calculate $cy + dx$. i.e., 35 + 24 = 59 and add the remainder 3 to it to make it 62, keep 3 in the answer and carry forward 6.

Step 4. Calculate $by + cx = 40 + 28 = 68$ and add the remainder 6 to it to make it 74, keep 4 in the answer and carry forward 7.

Step 5. Calculate $ay + bx = 25 + 32 = 57$ and add the remainder 7 to it to make it 64, keep 4 in the answer and carry forward 6.

Step 6. Calculate $ax = 20$ and add the remainder 6 to it to make it 26, which is the left-hand part of the answer.

Therefore, arranging the digits of the answer, we get 264420.

Example 2 : 3689 × 26

```
3  6  8  9
      2  6
----------
```

3 × 2/3 × 6 + 6 × 2/6 × 6 + 8 × 2/8 × 6 + 9 × 2 / 9 × 6

6 / 18 + 12/36 + 16/48 + 18/54

6 / 30 / 52 / 66 / 54 → adding carries to the previous digits,

6	30	52	71	4
3	30	59	1	4
6	35	9	1	4
9	5	9	1	4

So, 3689 × 26 = 95914

Example 3 : 8745 × 16

```
8 7 4 5
    1 6
-------
```

= 8/ 8 × 6 + 7 × 1 / 7 × 6 + 4 × 1 / 4 × 6 + 5 × 1 / 30

= 8 / 48 + 7 / 42 + 4/24 + 5 /30

= 8 /55/46/29/30 Arranging the digits,

=139920

Thus, 8745 × 16 = 139920.

Try these.

(1) 4053 × 26
(2) 5876 × 65
(3) 6879 × 67
(4) 9874 × 35
(5) 6312 × 49
(6) 9632 × 47
(7) 6987 × 48
(8) 6987 × 42
(9) 3658 × 47
(10) 6987 × 632

Answers :

1) 105378 2) 381940 3) 460893 4) 345590 5) 309288
6) 452704 7) 335376 8) 293454 9) 171926 10) 4415784

Multiplication of 3-digit numbers by 3-digit numbers

This can be explained with the help of the alphabets a,b,c and x,y,z.

```
          a   b   c
          x   y   z
-------------------
         az  bz  cz
     ay  by  cy
 ax  bx  cx
-------------------
```

ax /ay + bx/az +by+cx/bz + cy/cz

Example 1 : 566 × 789

5 6 6

7 8 9

= 5 × 7/5 × 8 + 6 × 7 / 5 × 9 + 7 × 6 + 6 × 8/6 × 9 + 8 × 6 /6 × 9

= 35 / 40 + 42 /45 + 42 + 48/54 + 48 /54

= 35 / 82 / 135/102 /54 Adding carries to the previous digits,

= 446574

Thus, 566 × 789 = 446574

Example 2 : 366 × 492

3 6 6

4 9 2

= 12/3 × 9 + 4 × 6/3 × 2 + 4 × 6 + 6 × 9/6 × 2 + 9 × 6/6 × 2

=12/27 + 24/ 6 + 24 + 54 / 12 + 54 / 12

=12/51/84/66/12 Adding carries to previous digits,

=180072

Thus, 366 × 492 = 180072

Example 3 : 7 8 2 × 4 9 8

7 8 2

4 9 8

= 28/7 × 9 + 4 × 8/ 7 × 8 + 4 × 2 + 8 × 9/8 × 8 + 9 × 2/2 × 8

= 28/63 + 32 / 56 + 8 + 72 / 64 + 18 / 16

= 28 / 95 / 136 / 82 / 16 adding carries to the previous digits,

= 389436

Thus, 782 × 498 = 389436.

Try these:

(1) 145 × 631 (2) 365 × 412

(3) 416 × 367 (4) 153 × 148

(5) 369 × 149

Answers :

1) 91495 2) 150380 3) 152672 4) 22644 5) 54981

Multiplication of 3 numbers by Vedic method

Case 1. When the numbers to be multiplied are nearer to the power of 10.

Example 1 : 13 × 14 × 15

Step 1. Write the deviation of each number from its base. Here base is 10.

Number Deviation from the base

13 +3

14 +4

15 +5

Step 2. Add all the deviations to the working base

10 + 3 + 4 + 5 = 22

Step 3. Make all the possible permutations of the deviation taking two at a time and add then.

3 × 4 + 3 × 5 + 4 × 5 = 12 + 15 + 20 = 47

Step 4. Multiple the deviations.

3 × 4 × 5 =60

Step 5. Arrange the results obtained in above steps as shown below.

22 / 47 / 60

Adding the carries to the previous digit, 2730

Thus, 1 3 × 14 × 15 = 2730.

Example 2 : 19 × 12 × 14

Step 1. Write the deviation of each number from its base.

19 +9

12 +2

14 +4

Step 2. Add all the deviations to the working base

$10 + 9 + 2 + 4 = 25$

Step 3. Make all the possible permutations of the deviation taking two at a time and add then.

$9 \times 2 + 2 \times 4 + 9 \times 4 = 18 + 8 + 36 = 62$

Step 4. Multiple the deviations

$9 \times 2 \times 4 = 72$

Step 5. Arrange the results obtained in above steps as shown below.

25 / 62 / 72

Adding the carries to the previous digit, answer becomes 3192.

Thus, $19 \times 12 \times 14 = 3192$.

Example 3 : $105 \times 115 \times 120$

Step 1. Write the deviation of each number from its base.

105 +5

115 +15

120 +20

Step 2. Add all the deviations with base. Here base is 100.

$100 + 5 + 15 + 20 = 140$

Step 3. Make all the possible permutations of the deviation taking two at a time and add then.

$5 \times 15 + 15 \times 20 + 5 \times 20 = 75 + 300 + 100 = 475$

Step 4. Multiple the deviations

$5 \times 15 \times 20 = 1500$

Step 5. Arrange the results obtained in above steps as shown below.

140/475/1500

Since the base is 100, take first two digits from each part and the remaining digit is to be carry forward to the previous parts.

Thus take 00 from first part, this is the first part of the answer. Carry forward 15 as remainder to previous part. Then 475 changes to 490. Take 90 from 490. 90 is the second part of the answer and carry forward 4 to 140. i.e., 144.

Answer of 105 × 115 × 120 is 1449000.

Example 4 : 1009 × 1010 × 992

Step 1. Write the deviation of each number from its base.

1009	+9
1010	+10
992	−8

Step 2. Add all the deviations with base. Here base is 1000.

1000 + 9 + 10 + −8 = 1011

Step 3. Make all the possible permutations of the deviation taking two at a time and add then.

9 × 10 + 10 × −8 + 9 × −8 = 90 + −80 + −72 = −62

Step 4. Multiple the deviations

9 × 10 × −8 = −720

Step 5. Arrange the results obtained in above steps as shown below.

1011 / −62 / −720

Step 6. We must change this negative number by replacing it with it's compliment. In this case base is 1000, so take the complement of 720. Complement of 720 is 1000 – 720 = 280.

→ 280 is the first part of the answer from right.

Whenever we do a replacement of this type, we must also take 1 as carry. 1 needs to be subtracted from the answer of previous part.

Then – 62 – 1 = – 63.

Step 7. As explained in step 6, We must convert −63 to its complement. Since the base is 1000, 1000 – 63 = 937. 937 is the second part of the answer.

Whenever we do a replacement of this type, we must also take 1 as carry. 1 needs to be subtracted from the answer of previous part. i.e., 1011 – 1 = 1010 which is the third part of the answer.

So, 1009 x 1010 x 992 is 1010937280

Example 5 : $992 \times 998 \times 980$

Step 1. Write the deviations of each number from its base.

992 – 8

998 – 2

980 – 20

Step 2. Add all the deviations with base. Here base is 1000.

$1000 + -8 + -2 + -20 = 970$

Step 3. Make all the possible permutations of the deviation taking two at a time and add then.

$-8 \times -2 + -2 \times -20 + -8 \times -20 = 16 + 40 + 160 = 216$

Step 4. Multiple the deviations

$-8 \times -2 \times -20 = -320$

Step 5. Arrange the results obtained in above steps as shown below.

970 / 216 / -320

Since 320 has minus sign, take the complement of 320. i.e., 1000 – 320 = 680. Since we have taken the complement, 1 has to be subtracted from the previous part. i.e., 216 – 1 = 215.

970 will remain as it is.

Answer is 970215680.

Try these:

(1) $9 \times 12 \times 13$

(2) $109 \times 98 \times 114$

(3) $95 \times 82 \times 102$

(4) $82 \times 110 \times 104$

(5) $998 \times 112 \times 983$

Answers :

1) !404 2) 1217748 3) 794580 4) 938080 5) 109875808

CASE 2 : When the working base is 50, 500, 5000, 50000

Step 1. Write the deviation of each number from its base.

Step 2. Add all the deviations with base. Here base is 100.

Step 3. Make all the possible permutations of the deviation taking two at a time and add then.

Step 4. Multiple the deviations

Step 5. Arrange the results obtained in above steps.

Step 6. Keep the right hand part of the result as it is. Divide the results of second part from right by 2.

Step 7. Divide the results of third part from right by 4.

Example 1. $58 \times 54 \times 52$

Step 1. Write the deviation of each number from its base.

58 2

54 4

52 8

Step 2. Add all the deviations with base. Here working base is 50.

$50 + 2 + 4 + 8 = 64$

Step 3. Make all the possible permutations of the deviation taking two at a time and add then.

$2 \times 4 + 4 \times 8 + 2 \times 8 = 8 + 32 + 16 = 56$

Step 4. Multiple the deviations

$2 \times 4 \times 8 = 64$

Step 5. Arrange the results obtained in above steps.

64 56 64

Step 6. Keep the right hand part of the result as it is. Divide the results of second part from right by 2. i.e., 56 /2 =28.

Step 7. Divide the results of third part from right by 4. i.e., 64 / 4 = 16

Here theoretical base is 100 and working base is 50. so ½ = 50 and ¼ = 25 is taken.

Arranging the answers, 16 28 64

Thus, $58 \times 54 \times 52 = 162864$

Multiplication by 11

Example 1 : 26×11

Step 1. Write the digit on left hand side of the number first. Here the number is 26, so 2 is written first.

Step 2. Add the two digits of the given number and write it in between. Here $2 + 6 = 8$

Step 3. Now write the second digit on extreme right. Here the digit is 6.

So, $26 \times 11 = 286$

Example 2 : 156×11

Step 1. Write the digits on left hand side of the number first. Here the number is 156. So, 1 is written first from left.

Step 2. First Add 1 and 5. i.e., $1 + 5 = 6$, this is the second digit. If any carry is getting in the next step, it has to be added to the result got in this step.

Step 3. Add 5 and 6. i.e., $5 + 6 = 11$. Third digit is 1. Take trailing 1 of 11 as carry and add the same to the previous digit. i.e., second digit 6 changes to 7.

Step 4. Write the digit on extreme right hand as the right hand digit of the answer. Here the digit is 6.

Arranging the digits got on above steps, 1716

So, 156×11 is 1716.

Example 3 : 2569×11

Step 1. First digit from left - 2 same as it is in the multiplicand

Step 2. Second digit, 2 + 5 = 7

Step 3. Third digit, 5 + 6 = 11

Step 4. Fourth digit, 6 + 9 = 15

Step 4. Fifth digit – 9 same as it is in the multiplicand

Arranging the answers

2 7 11 15 9 carry the remainders to previous digits,

= 2 7 + 11 + 159

= 28259

So, 2569 × 11 is 28259

Example 4 : 96587 × 11

Step 1. First digit from left - 9 same as it is in the multiplicand

Step 2. Second digit → 9 + 6 =15

Step 3. Third digit → 6 + 5 = 11

Step 4. Fourth digit → 5 + 8 = 13

Step 5. Fifth digit → 8 + 7 = 15

Step 6. Sixth digit → 7 same as it is in the multiplicand

Arranging the digits,

9 15 11 13 15 7 → adding the carry to the previous digits,

1062457 is the answer.

So, 96587 × 11 = 1062457.

Try these :

(1) 12 × 11 (2) 56 × 11

(3) 125 × 11 (4) 459 × 11

(5) 981 × 11 (6) 1012 × 11

(7) 1963 × 11 (8) 8951 × 11

(9) 12559 × 11 (10) 98521 × 11

Answers :

1) 132 2) 616 3) 1375 4) 5049 5) 10791

6) 11132 7) 21593 8) 98461 9) 138149 10) 1083731

Multiplication by 111

Example 1 : 145 × 111

Step 1. First digit of the multiplicand will be the first digit of the answer from left. 1 is the first digit of the answer.

Step 2. 1 + 4 = 5, 5 is the second digit of the answer from left.

Step 3. 1 + 4 + 5 = 10 (multiplying by 111, so three digits are added at a time). 10 is the third digit of the answer.

Step 4. 4 + 5 = 9 is the fourth digit of the answer.

Step 5. Last digit of the multiplier and last digit of the answer will be same. i.e., 5.

Arranging the answers,

1 5 10 9 5 adding carries to the previous digits,

1 6 0 9 5 is the answer.

So, 145 × 111 = 16095.

Example 2 : 859 x 111

Step 1. First digit from the left → 8

Step 2. Second digit → 8 + 5 = 13

Step 3. Third digit → 8 + 5 + 9 = 22

Step 4. Fourth digit → 5 + 9 = 14

Step 5. Fifth digit → 9

Arranging the numbers

8 13 22 14 9 adding the carries to the previous digits, we get 95349.

So, 859 × 111 = 95349.

Example 4 : 8965 × 111

Step 1. First digit is 8

Step 2. Second digit is 8 + 9 = 17

Step 3. Third digit is 8 + 9 + 6 = 23

Step 4. Fourth digit is 9 + 6 + 5 = 20

Step 5. Fifth digit is 6 + 5 = 11

Step 6. 6th digit is 5

Arranging the digits,

8 17 23 20 11 5 adding the carries to the previous digit, we get 995115.

So, 8965 × 111 is 995115.

Example 5 : 15695 × 111

Step 1. First digit is the left most digit of the multiplicand. i.e., 1

Step 2. Second digit is 1 + 5 = 6

Step 3. Third digit is 1 + 5 + 6 = 12

Step 4. Fourth digit is 5 + 6 + 9 = 20

Step 5. Fifth digit is 6 + 9 + 5 = 20

Step 6. Sixth digit is 9 + 5 = 14

Step 7. Seventh digit is the last digit of the multiplicand. i.e., 5.

Arranging the digits,

1 6 12 20 20 14 5 adding carries to the previous digits,

1742145 is the answer.

So, 15695 × 111 = 1742145.

Try these:

(1) 12 × 111
(2) 125 × 111
(3) 456 × 111
(4) 6985 × 111

(5) 7895 × 111 (6) 78496 × 111

(7) 12365 × 111 (8) 99 × 111

(9) 478596 × 111 (10) 781 × 111

Answers :

1) 1332 2) 13875 3) 50616 4) 775335 5) 876345

6) 8713056 7) 1372515 8) 10989 9) 53124156

10) 86691

Multiplication by 1111

Example 1 : 1012 × 1111

Step 1. First digit of the multiplicand is the first digit of the answer. i.e., 1 is the first digit from left is the first digit of the answer.

Step 2. Adding first 2 digit from left. i.e., 1 + 0 = 1 is the second digit.

Step 3. Adding first 3 digits from left. i.e., 1 + 0 + 1 = 2 is the third digit.

Step 4. Adding first 4 digits from left. i.e., 1 + 0 + 1 + 2 = 4 is the fourth digit.

Step 5. Adding last 3 digits from right 0 + 1 + 2 = 3 is the fifth digit.

Step 6. Adding last 2 digits from right i.e., 1 + 2 = 3 is the sixth digit.

Step 7. Last digit from left is the seventh digit of the answer. i.e., 2.

Arranging the digits,

1 1 2 4 3 3 2

So, 1012 × 1111 = 1124332.

Example 2 : 324 × 1111

Step 1. Since multiplier 9999 has 4 digits, put 0 before 324. Then number becomes 0324. First digit of the answer is the left most digit of 0324. i.e., 0.

Step 2. Adding first 2 digits from left, i.e., 0 + 3 = 3 is the second digit of the answer.

Step 3. Adding first 3 digits from left, i.e., 0 + 3 + 2 = 5 is the third digit of the answer.

Step 4. Adding first 4 digits from left, i.e., 0 + 3 + 2 + 4 = 9 is the fourth digit of the answer.

Step 5. Adding last 3 digits from right, i.e., 3 + 2 + 4 = 9 is the fifth digit of the answer.

Step 6. Adding last 2 digits from right i.e., 2 + 4 = 6 is the sixth digit of the answer.

Step 7. Last digit from left or first digit from right is the seventh (right most) digit of the answer. i.e., 4.

Arranging the digits, 0 3 5 9 9 6 4.

So, 324 × 1111 = 359964.

Example 3 : 45 × 1111

Step 1. Since the multiplier (here 1111) has 4 digits, append 00 digits before 45 making 45 to 0045.

Then first digit from left is 0.

Step 2. Second digit is 0 + 0 = 0.

Step 3. Third digit is 0 + 0 + 4 = 4.

Step 4. Fourth digit is 0 + 0 + 4 + 5 = 9.

Step 5. Fifth digit is the sum of first 3 digits from right i.e., 0 + 4 + 5 = 9.

Step 6. Sixth digit is the sum of first 2 digits from right i.e., 4 + 5 = 9.

Step 7. 5 is the right most digit of the answer.

Arranging the digits, 0049995.

So, 45 × 1111 is 49995.

Example 4 : 45689 × 1111

Step 1. Trailing 4 of 45689 is the left most digit of the answer.

Step 2. 4 + 5 = 9 is the second digit from left of the answer.

Step 3. 4 + 5 + 6 = 15 is the third digit from left of the answer.

Step 4. 4 + 5 + 6 + 8 = 23 is the 4th digit from left of the answer.

Step 5. Sum of last 4 digits from right, i.e., 5 + 6 + 8 + 9 = 28 is the 5th digit from left of the answer.

Step 6. 6 + 8 + 9 = 23 is the 6th digit from left of the answer.

Step 7. 8 + 9 = 17 is the 7th digit from left of the answer.

Step 8. 9 is the last digit of the answer.

Arranging the digits,

4 9 15 23 28 23 17 9

Adding carries to the previous digits, answer changes to 50760479.

Thus, 45689 × 1111 = 50760479.

Try these :

(1) 12 × 1111
(2) 456 × 1111
(3) 3698 × 1111
(4) 789 × 1111
(5) 978 × 1111
(6) 1951 × 1111
(7) 9847 × 1111
(8) 45621 × 1111
(9) 47856 × 1111
(10) 74156 × 1111

Answers :

1) 13332 2) 506616 3) 4108478 4) 876579 5) 1086558
6) 2167561 7) 10940017 8) 50684931 9) 53168016
10) 82387316

6

Division

Division

In the conventional procedure for division, the process is of the following form.

```
            Quotient
            ________
Divisor ) Dividend        or Divisor ) Dividend ( Quotient
          ----------                  ----------
          ----------                  ----------
          __________                  __________
          Remainder                   Remainder
```

But in the Vedic process, the format is

```
Divisor ) Dividend
          --------
          --------
_____________________
Quotient / Remainder
```

The conventional method is always the same irrespective of the divisor. But Vedic methods are different depending on the nature of the divisor.

Example 1 : Consider the division 2360 / 13

(i) Conventional method:

```
13) 2360 (181
    13
   ____
    106
    104
  -----------
     20
     13
  -----------
      7
```

Thus quotient = 181 and remainder = 7

The below methods are used in Vedic Mathematics for division.

Nikhilam Method

This method is useful when divisor is nearer and less than the base.

Example 1 : Divide 239 by 9.

Step 1. Here divisor is 9 and dividend is 239. Nearest power of divisor 9 is 10. Obtain the modified divisor (M.D) by applying the Nikhilam formula i.e., last from 10 and rest from 9. i.e., minus last digit from 10 and minus rest digits from 9.

Modified divisor is 1.

Step 2. Split dividend in 2 parts (Quotient & Remainder) in such a way that the remainder to have same number of digits as that of divisor. In this case it is 1 digit. Here 23 is quotient and 9 is remainder.

	9	2 3	9
M.D.	1		

Step 3. Multiply the M.D with first column digit of the dividend. Here it is 1. i.e., $2 \times 1 = 2$. Write this number in the second and third column. Here no value to put in the third column.

	9	2	3	9
M.D	1		2	

Step 4. Add the numbers in the column 2, here $3 + 2 = 5$. Multiply M.D 1 by 5 and put in third column. i.e.,

	9	23	9
M.D	1	2	5
		25	14

25 is the quotient and 14 is the remainder.

Remainder cannot be greater than divisor. So divide again 14 by 9, then, 1 is the quotient and 5 is the remainder.

Add the quotient got by the above method to the quotient got by Nikhilam Method.

Quotient changes to 26 and remainder changes to 5.

Example 2 : Divide 425 by 95.

Here divisor is 95 and dividend is 425.

Step 1. Obtain the modified divisor (M.D) by applying the Nikhilam formula i.e., last from 10 and rest from 9. i.e., minus last digit from 10 and minus rest digits from 9. i.e., modified divisor is 5.

Step 2. Split dividend in 2 parts (Quotient & Remainder) in such a way that the remainder to have same number of digits as that of divisor. In this case divisor 95 has 2 digit. Here 4 is quotient and 2 5 is remainder.

	95	4	25
M.D	5		

Step 3. Multiply the M.D with first column digit of the dividend. Here it is 4. i.e., 5 × 4 = 20. Write this number in the second column.

	95	4	2 5
M.D	5		2 0

Step 4. Add the numbers in column 2, here 25 + 20 = 45.

	95	4	2 5
M.D	5		2 0
		4	4 5

Now 4 is the quotient and 45 is the remainder.

Example 3 : Divide 1648 by 92.

Step 1. Obtain the modified divisor (M.D) by applying the Nikhilam formula i.e., last from 10 and rest from 9. i.e., minus last digit from 10 and minus rest digits from 9. i.e., modified divisor is 08.

Step 2. Split dividend in 2 parts (Quotient & Remainder) in such a way that the remainder to have same number of digits as that of divisor. In this case divisor 92 has 2 digit. Here 16 is quotient and 48 is remainder.

92	16	48
M.D. 08		

Step 3. Multiply the M.D with first column digit of the dividend. Here it is 4. i.e., 16 × 8 = 128. Write 128 in 2nd column.

92	16	48
M.D 08	0	128

Step 4. Add values of 2nd column here it is 128+48 = 176

92	16	4 8
M.D 08	0	1 2 8
	16	176

Now, 16 is the quotient and 176 is the remainder. Since remainder cannot be greater than divisor, divide the remainder by divisor again. Then we get 1 as quotient and 84 as remainder. Add the quotient 1 to earlier quotient 16.

So, 17 is the quotient and 84 is the remainder.

Example 4 : Divide 12986 by 994.

Step 1. Here modified divisor is 006(1000 – 994).

Step 2. Split dividend in 2 parts (Quotient & Remainder) in such a way that the remainder to have same number of digits as that of divisor. In this case divisor 994 has 3 digits. Then 12 is the quotient and 986 is the remainder.

994	1 2	9 8 6
M.D. 006		

Step 3. Multiply the M.D. with first column digit of the dividend. Here it is 12 × 006 = 72. Write 72 in 2nd, column.

994	1 2	9 8 6
M.D. 006		72

Step 4. Add values of 2nd column here it is 986+72=1058

994	1 2	9 8 6
M.D. 006	0	7 2

Adding the values in all the columns,

994	1 2	9 8 6
M.D. 006	0	7 2
	1 2	1058

Now 12 is the quotient and 1058 is the remainder. But remainder cannot be greater than divisor. Then divide remainder 1058 by 994. Then 1 gets as quotient 64 as remainder. Add this quotient 1 to the quotient got in step 4. i.e., 12 +1 = 13.

Now quotient is 13 and remainder is 64.

Try these :

(1) 195 / 9

(2) 658 / 91

(3) 1298 / 993

(4) 16893 / 997

(5) 48631 / 9997

Answer :

1) Q - 21, R - 6 2) Q - 7, R - 21 3) Q - 1, R - 305

4) Q - 16, R - 941 5) Q - 4, R - 8643

Paravartya Sutra:

This is another Shortcut Method of Division in Vedic Mathematics.

Paravartya Sutra is a Specific Method for division in Vedic Maths. This method is applicable only when the divisor is nearer to the base 10 or the power of 10. This sutra works effectively when the first digit of the divisor is 1.

Using Paravartya Sutra, you can easily divide when divisor is like 11, 113, 1104 etc.

This sutra is similar to the Nikhilam sutra except for the following:

(a) In the divisor column, write the complement of divisor from the base and also the complement with change of sign. Thus + becomes – and conversely.

(b) Complement = Divisor – Base

(c) Here the multiplication of quotient digits is performed with the complement, having changed sign for each digit.

Example 1 : Divide 198 by 13.

Step 1. Here base is 10. Complement = 13 – 10 = 3. Complement with changed sign = –3.

Step 2. Arrange the dividend 198 in 2 columns (quotient and remainder) such a way that remainder contains 1 digit number. Split 198 like 19 in first column(quotient) and 8 in the second column(Remainder).

	13	19	8
Complement	3		
Revised Complement =	– 3		

Step 3. Multiply 1st digit of the quotient part 1 with revised complement i.e., 1 × –3 = –3. Write down -3 below the second digit of the quotient part as below.

	13	19	8
Complement	3	–3	
Revised Complement =	– 3		

Step 4. Add the digits 9 + –3 and multiply the same with revised complement and write the answer in the next column. i.e., 9 + –3 = 6, multiply 6 by –3 = –18, write –18 below 8.

	13	19	8
Complement	3	−3	
Revised Complement =	− 3		−18

Step 5. Adding the digits by column wise,

	13	19	8
Complement	3	−3	−18
Revised Complement =	− 3		
		16	−10

Step 6. Remainder cannot be written in negative terms,

198 = 13 × 16 + Remainder. It can also be written

198 = (13 × 15 + 13) + Remainder.

So, quotient 16 can be written as 15 + 13. Add −10 with 13, then remainder is changed to 3 and quotient is changed to 15.

So Dividing 198 by 13, 15 is the quotient and 3 is the remainder.

Example 2 : Divide 468 by 101.

Step 1. Here base is 100. Complement = 101 – 100 = 1. Complement with changed sign = −1. Since base is 100, append a 0 before the complement −1.

Step 2. Arrange the dividend 468 in 2 columns (quotient and remainder) such a way that remainder contains 2 digit number. Split 468 like 4 in first column(quotient) and 68 in the second column(Remainder).

	101	4	68
Complement	01		
Revised Complement =	− 01		

Step 3. Multiply 1st digit of the quotient 4 with revised complement i.e., 4 × −01 = −04. Write down −0 – 4 below the second and third digit of the quotient part as below.

	101	4	6	8
Complement	01		–0	–4
Revised Complement =	– 01			

Step 4. Adding digits by column wise,

	101	4	6	8
Complement	01		–0	–4
Revised Complement =	– 01			
		4	64	

So, we get 4 as quotient and 64 as remainder.

Example 3 : Divide 1298 by 108.

Step 1. Here base is 100. Complement = 108 – 100 = 8. Revised complement with changed sign = –08

Step 2. Arrange the dividend 1298 in 2 columns (quotient and remainder) such a way that remainder contains 2 digit number as base is 100. Split 1298 like 12 in first column(quotient) and 98 in the second column(Remainder).

	108	12	9 8
Complement	8		
Revised Complement =	–0 –8		

Step 3. Multiply first digit of the quotient 1 by revised complement. i.e., 1 × –0 –8 = –0 –8.

Write down –0 – 8 below 2nd and 3rd digits.

	108	1 2	9	8
Complement	8	–0	– 8	
Revised Complement =	–0 –8			

Step 4. Add the digits, 2 + –0 = 2, multiply 2 by –0 –8, we get –1 – 6, write down the –1 – 6 below the digits 9 and 8 in the remainder column.

108	1 2	9 8
Complement 8	−0	− 8
Revised Complement = −0 −8		−1 −6

Step 5. Add the digits by column wise, we get 1 2 in quotient column and 02 in the remainder column.

So, Answer is : Quotient = 12 and remainder = 2.

Example 4 : Divide 16192 by 1012.

Step 1. Here base is 1000. Complement = 1012 −1000 = 12. Revised complement with changed sign = − 1 −2

Step 2. Arrange the dividend 16192 in 2 columns (quotient and remainder) such a way that remainder contains 3 digit number as base is 1000. Split 16192 like 16 in first column(quotient) and 192 in the second column(Remainder).

1012	16	192
Complement 12		
Revised Complement = − 1 −2		

Step 3. Multiply first digit of the quotient 1 by revised complement. i.e., 1 × −1 −2 = −1 −2

Write down 0 −1 −2 below 2nd, 3rd and 4th digits.

1012	1 6	1 9 2
Complement 12	−0	−1−2
Revised Complement = − 0 − 1 −2		

Step 4. Add the digits, 6 + −0 = 6, multiply 6 by −0 −1 − 2, we get −6 − 12. −6 −12 can be written as −7 − 2 Write down the below the digits 9 and 2 in the remainder column.

1012	1 6	1 9 2
Complement 12	−0	−1 −2
Revised Complement = − 0 − 1 −2		−0 −7 −2

Step 5. Adding the digits column wise, 16 0 0 0.

So, 16 is the quotient and 0 is the remainder.

Try these.

(1) 802 / 15

(2) 925 / 105

(3) 4569 / 16

(4) 95687 / 107

(5) 987454 / 1011

Answer :

1) Q - 53, R - 7 2) Q - 8, R - 85 3) Q - 285, R - 9

4) Q - 894, R - 29 5) Q - 976, R - 718

7

Square and Square Roots

Squaring and square roots

Square of a number is that number, multiplied by itself.

Square of numbers ending in 5

In Vedic Mathematics Sutra : 'By one more than previous one" can be used for finding the square of a number ending in 5.

Square of a number ending in 5 = [left digit(s) ×(Left digit(s) +1)] 25

Example 1 : 15 × 15

For the number last digit is 5 and previous digit is 1.

One more than the previous one, i.e., 1 +1 =2.

As per the method, multiply the previous digit by one more than itself. i.e., 1 × 2 = 2.

This is the left hand side of the answer.

Right hand side of the answer is 5 × 5 = 25.

Thus 15 × 15 is 225.

Example 2 : 35 × 35

For the number last digit is 5 and previous digit is 3.

One more than the previous one, i.e., 3 + 1 = 4. As per the method, multiply the previous digit by one more than itself. i.e., 3 × 4 = 12. This is the left hand side of the answer.

Right hand side of the answer is 5 × 5 = 25.

Thus 35 × 35 is 1225.

Example 3 : 105 × 105

For the number last digit is 5 and previous digit is 10.

One more than the previous one, i.e., 10 + 1 = 11. As per the method, multiply the previous digit by one more than itself. i.e., 10 × 11 = 110. This is the left hand side of the answer.

Right hand side of the answer is 5 × 5 = 25.

Thus 105 × 105 is 11025

Example 4 : 125 × 125

For the number last digit is 5 and previous digit is 12.

One more than the previous one, i.e., 12 + 1 = 13. As per the method, multiply the previous digit by one more than itself. i.e., 12 × 13 = 156. This is the left hand side of the answer.

Right hand side of the answer is 5 × 5 = 25.

Thus 125 × 125 is 15625

Example 5 : 195 × 195

For the number last digit is 5 and previous digit is 19.

One more than the previous one, i.e., 19 + 1 = 20. As per the method, multiply the previous digit by one more than itself. i.e., 19 × 20 = 380. This is the left hand side of the answer.

Right hand side of the answer is 5 × 5 = 25.

Thus 195 × 195 is 38025

Example 6 : 1025 × 1025

For the number last digit is 5 and previous digit is 102.

One more than the previous one, i.e., 102 + 1 = 103. As per the method, multiply the previous digit by one more than itself. i.e., 102 × 103. This is the left hand side of the answer.

Right hand side of the answer is 5 × 5 = 25.

Thus 1025 × 1025 = 1050625

Example 7 : 77.5 × 77.5

For the number last digit is. 5 and previous digit is 77.

One more than the previous one, i.e., 77 + 1 = 78. As per the method, multiply the previous digit by one more than itself. i.e., 77 × 78 = 6006. This is the left hand side of the answer.

Right hand side of the answer is. 5 ×.5 = .25.

Thus 77.5 × 77.5 = 6006.25

Try these.

(1) 45 × 45	(2) 95 × 95
(3) 115 × 115	(4) 135 × 135
(5) 175 × 175	(6) 235 × 235
(7) 1035 × 1035	(8) 2015 × 2015
(9) 4025 × 4025	(10) 8505 × 8505

Answers :

1) 2025 2) 9025 3) 13225 4) 18225 5) 30625

6) 55225 7) 1071225 8) 4060225 9) 16200625

10) 72335025

Square of any two digit numbers

The sub-sutra Anurupsyena helps to find out the square of any two digit number.

Step 1. Square the digit at tenth place. This is the first column of the answer from left.

Step 2. Multiply digits of unit place and tenth place. Take the double of the same.

Step 3. Square the digit at ones place.

Arrange the numbers adding carries to the previous digits.

Example 1 : 57 × 57

Step 1. Multiply 5 by 5 = 25

Step 2. Multiply 5 by 7 = 35. Take the double. i.e., 2 × 35 = 70

Step 3. 7 × 7 = 49

Arranging the digits,

25 70 49

Adding the carries to the previous digits, answer changes to 3249.

So, $57 \times 57 = 3249$

Example 2 : 58×58

Step 1. Multiply $5 \times 5 = 25$

Step 2. Multiply $5 \times 8 = 40$. Take the double i.e., $40 \times 2 = 80$

Step 3. Multiply $8 \times 8 = 64$

Arranging the digits,

25 80 64

Adding carried to the previous digits, answer changes to 3364

So, $58 \times 58 = 3364$.

Example 3 : 95×95

Step 1. $9 \times 9 = 81$

Step 2. $2 \times (9 \times 5) = 90$

Step 3. $5 \times 5 = 25$

Arrange the numbers

81 90 25

Adding the carries to the previous digits, answer changes to 9025.

So, $95 \times 95 = 9025$.

Try these :

(1) 24×24 (2) 36×36

(3) 78×78 (4) 96×96

(5) 23×23 (6) 64×64

(7) 37×37 (8) 49×49

(9) 58×58 (10) 84×84

Answers :

1) 576 2) 1296 3) 6084 4) 9216 5) 529 6) 4096

7) 1369 8) 2401 9) 3364 10) 7056

Squaring numbers near 50

Example 1 : 47 × 47

Step 1. 47 is 3 less than 50. First part of answer calculated as: 5 × 5 – 3 = 25 – 3 = 22.

Step 2. Second part is calculated as : 3 × 3 = 9. (need 2 digit number. So 9 can be written as 09).

Thus, 47 × 47= 2209

Example 2 : 42 × 42

Step 1. 42 is 8 less than 50. First part of the answer is

5 × 5 – 8 = 25 – 8 = 17

Step 2. Second part is 8 × 8 = 64.

Thus, 42 × 42 = 1764

Example 3 : 38 × 38

Step 1. 38 is 12 less than 50. First part of the answer is

5 × 5 – 12 = 25 – 12 = 13

Step 2. Second part is 12 × 12 = 144. Need only digit of 144, so leading 1 of 144 is carried over to first part of the answer. i.e., 13 + 1 = 14.

So, 38 × 38 = 1444

Example 4 : 61 × 61

Step 1. 61 is 11 more than 50. First part of the answer is

5 × 5 + 11 = 36

Step 2. 11 × 11 = 121. Trailing 1 of 121 is carried over to first part. Then first part is 37 and second part is 21

So, 61 × 61 = 3721

Squaring numbers near base :

Example 1 : 13 × 13

Step 1. The nearest base to 13 is 10. Deviation from base = 13 – 10 = 3. Add 3 to 13. i.e., 13 + 3 = 16. This is the first part.

Step 2. Take square of 3. i.e., 3 × 3 = 9 is the second part.

Arranging the digits,
169 is the answer.

So, 13 × 13 = 169.

Example 2 : 18 × 18

Step 1. The nearest base to 18 is 10. Deviation from base = 18 – 10 = 8. Add 18 to 8. i.e., 18 + 8 = 26. 26 is the first part.

Step 2. Take square of 8 i.e., 8 × 8 = 64 is the second part. Since base is 10, carry trailing 6 of 64 as a carry to previous digit. So, first part of the answer changes to 26 + 6 = 32. Second part is 4.

So, 18 × 18 = 324

Example 3 : 105 × 105

Step 1. The nearest base to 105 is 100. Deviation from base = 105 – 100 = 5. Add 5 to 105. i.e., 105 + 5 = 110. First part of the answer is 110.

Step 2. Second part is 5 × 5 = 25

Arranging the digits,

110 25

Since base 100 and 25 has 2 digits, no carry to add.

So, 105 × 105 = 11025.

Example 3 : 135 × 135

Step 1. Nearest base to 135 is 100. Deviation from base is 35. Add 135 to 35. 135 + 35 = 170 is the first part of the answer.

Step 2. 35 × 35 = 1225.

Arranging the numbers,

170 1225

Since base is 100, take 25 of 1225 as the second part and add trailing 12 of 1225 as carry to 170. The first part changes to 170 +12 = 182.

So, 135 × 135 = 18225

Example 4 : 1015 × 1015

Step 1. Nearest base to 1015 is 1000. Deviation from base is 15. Then 1015 + 15 = 1030.

Step 2. 15 × 15 = 225

Arranging the digits,

1030 225

So, 1015 × 1015 = 1030225.

Squaring numbers near sub - base:

Example 1 : 34 × 34

Step 1. 34 can be written as (3 × 30) + 4. Hence 30 is taken as the working base.

Deviation from the base 34 – 30 = 4. First part of the answer is 3(34 + 4) = 3 × 38 = 114.

Step 2. Second part of the answer is 4 × 4 = 16. Since base is 10, trailing 1 of 16 is carried over to first part. First part of the answer changes to 114 + 1 = 115.

So, 34 × 34 = 1156.

Example 2 : 46 × 46

Step 1. 46 can be written as (4 × 40) + 6. Hence 40 is taken as the working base.

Deviation from the base 46 – 40 = 6. First part of the answer is 4(46 + 6) = 4 × 52 = 208.

Step 2. Second part of the answer is 6 × 6 = 36. Since base is 10, trailing 3 of 36 is carried over to first part. First part of the answer changes to 208 + 3 = 211.

So, 34 × 34 = 2116.

Example 3 : 128 × 128

Step 1. 128 can be written as (1 × 100) + 28. Here 100 is taking as the working base. Deviation from working base is 128 – 100 = 28. First part of the answer is 1 × (128 + 28) = 156.

Step 2. Second part of the answer is 28 × 28 = 784. since base is 100, trailing 7 of 784 is carried over to first part. So, first part changes to 156 + 7 = 163. Second part of the answer is 84.

Answer is 16384.

Example 4 : 306 × 306

Step 1. 306 can be written as (3 × 100) + 28. Here 100 is taking as the working base. Deviation from working base is 306 – 300 = 6. First part of the answer is 3 × (306 + 6) = 3 × 312 = 936.

Step 2. Second part of the answer is 6 × 6 = 36.

Answer is 93636.

Example 5 : 4050 × 4050

Step 1. 4050 can be written as (4 × 1000) + 50. Here 1000 is taking as the working base. Deviation from working base is 4050 – 4000. First part of the answer is 4(4050 + 50) = 4 × 4100 = 16400.

Step 2. Second part of the answer is 50 × 50 = 2500. Since base is 1000, trailing 2 of 2500 is carried over to first part. Now first part changes to 16402. Second part is 500.

Answer is 16402500.

Try these.

(1) 118 × 118 (2) 56 × 56

(3) 19 × 19 (4) 134 × 134

(5) 211 × 211 (6) 409 × 409

(7) 805 × 805 (8) 5011 × 5011

(9) 10003 × 10003 (10) 20005 × 20005

Answers :

1) 13924 2) 3136 3) 361 4) 17956 5) 44521

6) 167281 7) 648025 8) 25110121 9) 100060009
10) 400200025

Squaring using Duplex method

Duplex method is one of the best squaring method in vedic Mathematics.

Before doing exercise in this method, we should know how to find out the duplex of numbers and should learn grouping of numbers.

Duplex of 1 digit number = Square of that number

Duplex of 3 = $3^2 = 9$

Duplex of 2 digit number = 2 × (product of digits)

Duplex of 34 = $2 \times (3 \times 4) = 2 \times 12 = 24$

Duplex of 3 digit number = 2 × (1st digit × 3rd digit) + (square of middle digit)

Duplex of 245 = $2(2 \times 5) + 4 \times 4 = 20 + 16 = 36$

Duplex of 4 digit number = 2 × (1st digit × 4th digit) + 2 × (2nd digit × 3rd digit)

Duplex of 1245 = $2(1 \times 5) + 2(2 \times 4) = 10 + 16 = 26$

Duplex of 5 digit number = 2 × (1st digit × 5th digit) + 2 × (2nd × 4th digit) + $(\text{middle digit})^2$

Duplex of 12345 = $2 \times (1 \times 5) + 2(2 \times 4) + 9 = 10 + 16 + 9 = 35$

The following pattern will help you to grouping of numbers.

$(11)^2 = 1\ 2\ 1$

$(111)^2 = 1\ 2\ 3\ 2\ 1$

$(1111)^2 = 1\ 2\ 3\ 4\ 3\ 2\ 1$

$(11111)^2 = 1\ 2\ 3\ 4\ 5\ 4\ 3\ 2\ 1$

Example 1 : Find the square of 26 using duplex method.

26 can be grouped as 2 26 6

Duplex of 2 = $2 \times 2 = 4$

Duplex of 26 = $2(2 \times 6) = 24$

Duplex of 6 = 6 × 6 = 36

Result can be written as 4 24 36. Trailing digit of every part is carried over to previous digits. Thus trailing digit 3 of 36 is carried over to 24. thus 24 is changed to 27.

Trailing digit 2 of 27 is carried over to 4. Thus 4 is changed to 6.

So, 26 × 26 = 676.

Example 2. Find out the square of 125 using duplex method.

125 can be grouped as 1 12 125 25 5

Duplex of 1 = 1 × 1 = 1

Duplex of 12 = 2(1 × 2) = 4

Duplex of 125 = 2(1 × 5) + 4 = 14

Duplex of 25 = 2(2 × 5) = 20

Duplex of 5 = 25

We get, 1 4 14 20 25. Carrying the leading digits to previous digits,

1 5 6 2 5

So, 125 × 125 = 15625.

Example 3 : Find out the square of 1365 using duplex method.

1365 can be grouped as 1 13 136 1365 365 65 5

Duplex of 1 = 1

Duplex of 13 = 2(1 × 3) = 6

Duplex of 136 = 2(1 × 6) + 9 = 21

Duplex of 1365 = 2(1 × 5) + 2(3 × 6) = 10 + 36 = 46

Duplex of 365 = 2(3 × 5) + 36 = 30 + 36 = 66

Duplex of 36 = 2(6 × 5) = 60

Duplex of 5 = 25

Result can be written as

1 6 21 46 66 60 25 (trailing 2 of 25 is carry forward to 60,

	making 60 to 62)
1 6 21 46 66 62 5	(trailing 6 of 62 is carry forward to 66 making 66 to 72)
1 6 21 46 72 2 5	(trailing 7 of 72 is carry forward to 46 making 46 to 53)
1 6 21 53 2 2 5	(trailing 5 of 53 is carry forward to 21, making 21 to 26)
1 6 26 3 2 2 5	(trailing 2 of 26 is carry forward to 6, making 6 to 8)
1 8 6 3 2 2 5	

Thus answer is 18 6 3 2 2 5

Try these :

(1) 154 × 154 (2) 235 × 235

(3) 479 × 479 (4) 1258 × 1258

(5) 3214 × 3214

Answers :

1) 23716 2) 55225 3) 229441 4) 1582564 5) 10329796

SQUARE ROOTS

Let us see some basics rules for finding square root.

The given number is first arranged in two-digit groups from right to left. For example, 5678 is arranged as 56 – one group and 78 is the other group. If the number is 345, this can be grouped like – 3 is one group and 45 is the other group.

The number of digits in the square root will be the same as the number of groups. For example, square root of 5678 will have 2 digits as it has two groups.

The squares of the first nine natural numbers end with 1, 4, 5, 6, 9, 0. This means

- If the unit digit of any number is 2, 3, 7 or 8, it do not have a perfect square root.
- If an exact square ends in 1, its square root ends in 1 or 9.
- If an exact square ends in 4, its square root ends in 2 or 8.
- If an exact square ends in 5, its square root ends in 5.
- If an exact square ends in 6, its square root ends in 4 or 6.
- If an exact square ends in 9, its square root ends in 3 or 7.
- If a perfect square is an odd number, the square root is also an odd number.
- If a perfect square is an even number, the square root is also an even number.
- A whole number, which ends with an odd numbers of 0's, can never be the square of a whole number.
- An exact square never ends in a 6 if the penultimate digit (digit that is next to the last digit) is even (eg. exact squares can not end in 26, 46, 86, etc.).
- An exact square never has an odd penultimate digit unless the final digit is a 6 (thus, exact squares can not end in 39,71, etc.).
- An exact square never ends with an even number when the last two digits taken together are not divisible by 4 (thus, no exact square can end in 22, 34 and other non-multiples of 4 if the last digit is even).

Vedic method for Square Root

The best method for finding the square root is Dwanda – Yoga or Duplex method.

The rules for finding square root can be explained as below.

Step 1. The given number is first arranged in two-digit groups from right to left. If on left hand side, a single digit is left, that will also be counted as a group.

Step 2. The number of digits in the square root will be the same as the number of groups derived from the given number.

Example 1 : Find the square root of 625.

Step 1. Arrange 625 from right in two digit groups. Arrange 6 as one group and 25 as other group.

Step 2. $2^2 < 6 < 3^2$.. First digit of the square root will be 2.

Step 3. Arrange 625 as below. Put the first digit of the square root 2 in the quotient row, Square the first digit of the quotient and subtract the same from 6. Double the quotient and set down as divisor.

Divisor 4	6	25
Quotient Q	2	

Remainder = $6 - 2^2 = 2$

Put the remainder below the next dividend 2. Hence the next dividend is 22.

Divisor 4	6	$_2$25
Quotient Q	2	

Step 4. Divide 22 by 4 put the quotient 5 in the quotient column.

Divisor 4	6	$_2$25
Quotient Q	2	5

Remainder = 22 – 20 = 2. Put the remainder 2 next below dividend 5 as below.

Step 5.

Divisor 4	6	$_2$2 $_2$5
Quotient Q	2	5

Next Dividend is 25. Subtract the square of new divisor 5 from dividend i.e., 25 – 25 = 0.

Since no more digits are left, square root of 625 is 25.

Example 2 : Find the square root of 1764.

Step 1. Arrange 1764 in two digits groups like 17 as one group and 64 as other group.

Step 2. $4^2 < 17 < 5^2$ First digit of the square root will be 4.

Step 3. Arrange 1764 as below. Put the first value of the square root 4 in the quotient row. Square the quotient 4 and subtract from 17, that is the remainder. Double the quotient and set down as divisor.

Divisor 8	17	64
Quotient	4	

Remainder = $17 - 4^2 = 1$

Step 4. Put the remainder 1 below the next dividend digit. Next dividend is 16.

Divisor 8	17	164
Quotient	4	

Step 5. Divide 16 by 8. Put the quotient 2 in the quotient column. Now remainder is left.

Divisor 8	17	$_1$64
Quotient	4	2

Step 6. Subtract the square of the new quotient from 4. i.e., 4 – 4 = 0

Since no more digits are left, square root of 1764 is 42.

Try these :

(1) 784 (2) 961

(3) 2304 (4) 3364

(5) 8100

Answers :

1) 28 2) 31 3) 48 4) 58 5) 90

8

Cube and Cube Roots

In Vedic Mathematics, tricks to find out the cube of a number have classified as Specific and General Methods. Specific Method can be applied when numbers satisfy certain conditions like numbers closer to power of 10. While General Method can be applied to any types of numbers.

Depending on Specific and General Techniques, Cubes in Vedic Mathematics are classified in the form of Sutras as below.

1. Anurupyena Sutra (Specific Technique)
2. Yavadunam Sutra (Specific Technique)

Anurupyena Sutra – Shorcuts to find cubes using Vedic Mathematics

Anurupyena Sutra is shortcut to find cube in Vedic Mathematics. It is general technique which can be applied to any 2 digit number.

$(a+b)^3 = a^3 + 3a^2b + 3ab^2 + b^3$. This can be written as below.

$(a+b)^3 = a^3 + a^2b + ab^2 + b^3$

$2a^2b + 2ab^2$

Example 1 : Find 15^3

To find the cube of root of 15, as per Vedic Mathematics, write 15^3 as $(1+5)^3$

$(a+b)^3 = a^3 + a^2b + ab^2 + b^3$

$2a^2b + 2ab^2$

Step 1. $1^3 + 1^2 \times 5 + 1 \times 5^2 + 5^3$

$2 \times 1 \times 5 + 2 \times 1 \times 25$

Step 2. 1 + 5 + 25 + 125

10 + 50

----

1 15 75 125 → Adding carries to the previous digits, (add trailing 12 of 125 to 75 making 75 to 87, then add the trailing 8 of 87 to 15 making 15 to 23, then add 2 of 23 to 1 making 1 to 3.)We get,

3 3 7 5

So, $15^3 = 3375$.

Example 2 : Find 23^3

To find the cube root of 23, as per Vedic Mathematics, $23^3 = (2 + 3)^3$

Step 1. $2^3 + 2^2 \times 3 + 2 \times 3^2 + 3^3$

$2 \times 2^2 \times 3 + 2 \times 2 \times 3^2$

Step 2. 8 + 12 + 18 + 27

24 + 36

----

8 36 54 27 → Adding carries to previous digits, we get

8 36 56 7

8 41 6 7

12 1 6 7 is the final answer.

So, $23^3 = 12167$.

Example 3 : Find 54^3

To find the cube root of 54, as per Vedic Mathematics method, $54^3 = (5 + 4)^3$

Step 1. $5^3 + 5^2 \times 4 + 5 \times 4^2 + 4^3$

$2 \times 5^2 \times 4 + 2 \times 5 \times 4^2$

Step 2. 125 + 100 + 80 + 64

200 + 160

----

125 300 240 64 → Adding carries to the previous digits,

125 300 246 4

125 324 6 4

157 4 6 4 is the final answer.

So, $54^3 = 157464$.

Example 4 : Find 73^3

As per Vedic Mathematics method, $73^3 = (7 + 3)^3$. This can be written like,

Step 1. $7^3 + 7^2 \times 3 + 7 \times 3^2 + 3^3$

$2 \times 7^2 \times 3 + 2 \times 7 \times 3^2$

Step 2. 343 147 63 27

294 126

----

343 441 189 27 Adding carries to the previous digits,

343 441 191 7

343 460 1 7

389 0 1 7 is the final answer.

So, $73^3 = 389017$.

Example 5 : Find 98^3.

For finding cube root of 98, 98^3 can be written as $(9 + 8)^3$

Step 1. $9^3 + 9^2 \times 8 + 9 \times 8^2 + 8^3$

$2 \times 9^2 \times 8 + 2 \times 9 \times 8^2$

Step 2.	729	648	576	512
		1296	1152	
	729	1944	1728	512

Adding carries to the previous digits,

729	1944	1779	2
729	2121	9	2
941	1	9	2 is the final answer.

So, $98^3 = 941192$.

Try these:

(1) 18^3 (2) 29^3

(3) 36^3 (4) 45^3

(5) 52^3 (6) 68^3

(7) 81^3 (8) 79^3

(9) 82^3 (10) 96^3

Answers :

1) 5832 2) 24389 3) 46656 4) 91125 5) 140608

6) 314432 7) 531441 8) 493039 9) 551368 10) 884736

Nikhilam Sutra

Nikhilam Sutra is more effective when the number is nearer to the base 10 or the power of 10.

The procedure for finding the cube of a number can be explained as

Step 1. Take the deviation of the number to be cubed from its base.

Step 2. Number to be cubed + (2 × deviation from the base × (sub base)^2)

Step 3. (3 × (deviation)^2) × sub base

Step 4. (Deviation)^3

Example 1 : Find the cube of 19 by Nikhilam Sutra.

Step 1. Here base is 10. Deviation from base is 9. In this case sub base is 1.

Step 2. (Number to be cubed + 2 × deviation from the base) × (sub base)^2 = (19 + 2 × 9) × 1 = 37.

Step 3. (3 x (deviation)^2) × sub base = 3 × 9^2 × 1 = 243.

Step 4. (Deviation)^3 = 9 ^{3} = 729.

Arranging the answers got in the above steps,

37	243	729	Adding carries to the previous digits, since base is 10, keep first digit and add the trailing digits to the previous digits,
37	315	9	
68	5	9	is the answer.

So, 19^3 = 6859.

Example 2 : Find the cube of 95.

Step 1. Here base is 100. Deviation from base is 95 – 100 = –5. sub base is 1.

Step 2. (Number to be cubed + 2 × deviation from the base) × (sub base)^2 = (95 + 2 × –5) × 1^2 = 95 + – 10 = 85.

Step 3. (3 × (deviation)^2) × sub base = 3 × (–5)2 × 1 = 75.

Step 4. (Deviation)^3 = (–5)^3 = –125.

Arranging the answers got in the above steps,

85	75	–125	We cannot write -125, So, 75 is the 100th place digit, take 100 from 75 and add to -125
85	74	100 –125	
85	74	– 25	
85	73	100 – 25	
85	73	75	

Sine the base is 100, so there should be 2 digits in each digit separator.

So, $95^3 = 857375$.

Example 3 : Find the cube of 110.

Step 1. Base is 100. Deviation from the base is 110 – 100 = 10. Sub base is 1.

Step 2. (Number to be cubed + 2 × deviation from the base) × $(\text{sub base})^2 = (110 + 2 \times 10) \times 1 = 130$.

Step 3. $(3 \times (\text{deviation})^2) \times \text{sub base} = 3 \times 10^2 \times 1 = 300$.

Step 4. $(\text{Deviation})^3 = 10^3 = 1000$.

Arranging the digits,

130	300	1000	Adding the carries to previous digits keeping 2 digits in each digit separator,
130	310	00	
133	10	00	is the answer.

So, $110^3 = 1331000$.

Example 4 : 306^3

Step 1. 306 = 3 × 100 + 6. Here base is 100 and sub base is 3. Deviation from base is 6.

Step 2. (Number to be cubed + 2 x deviation from the base) x $(\text{sub base})^2 = (306 + 2 \times 6) \times 3^2 = 2862$.

Step 3. $(3 \times (\text{deviation})^2) \times \text{sub base}$.

$$3 \times 6^2 \times 3 = 324$$

Step 4. $(\text{Deviation})^3 = 6^3 = 216$.

Arranging the digits,

2862	324	216	Adding carries to previous digits keeping 2 digits in each digit separator,
2862	326	16	

2865 26 16 is the answer.

So, $306^3 = 28652616$.

Example 5 : 1005^3

Step 1. Here base is 1000. Deviation from base is 5. Sub base =1.

Step 2. (Number to be cubed + 2 × deviation from base) × (sub base)^2 = $(1005 + 2 \times 5) \times 1^2 = 1015$.

Step 2. $(3 \times (\text{deviation})^2) \times \text{sub base} = 3 \times 5^2 \times 1 = 75$.

Step 3. $(\text{Deviation})^3 = 5^3 = 125$.

Arranging the digits,

1015 75 125 Since base is 1000, so there should be 3 digits in each digit separator). Since 75 has only two digits, append 0 to make it 3 digit.

1015 075 125 is the answer.

$(1005)^3 = 1015075125$.

Try these: Find the cube of the below numbers:

(1) 45 (2) 33

(3) 89 (4) 169

(5) 289 (6) 596

(7) 101 (8) 141

(9) 202 (10) 308

Answers :

1) 91125 2) 35937 3) 704969 4) 4826809 5) 24137569

6) 211708736 7) 1030301 8) 2803221 9) 8242408

10) 29218112

Yavadunam Sutra

This sutra is also applicable in cubing a number. Steps for yavadunam can be summarized as below.

Step 1. Number + (2 × excess / deficit from base).

Step 2. New excess / deficit × original excess / deficit.

Step 3. Cube of the original excess / deficit.

Example 1 : Find the cube of 16 by Yavadunam Sutra.

Step 1. Nearest base is 10. Excess from base = 16 – 10 = 6.

1st part = Number + (2 × Excess) = 16 + 2 × 6 = 28

Step 2. New excess × original excess.

(28 – 10) × 6 = 18 × 6 = 108

Step 3. Cube of the original excess = 6^3 = 216.

Arranging the digits,

28 108 216	Since base is 10, so there should be 1 digits in each digit separator). Add the trailing digits as carry.
28 129 6	
40 9 6	is the answer.

So, 16^3 = 4096.

Example 2 : Find the cube of 106 by Yavadunam Sutra.

Step 1. Nearest base for 106 is 100. Excess from base = 6.

1st part = Number + (2 × Excess) = 106 + (2 × 6) = 106 + 12 = 118

Step 2. New excess x Original excess = 18 × 6 = 108.

Step 3. Cube of excess = 6^3 = 216.

Arranging the numbers,

118 108 216	Since base is 100, so there should be 2 digits in each digit separator). Add the trailing digits as carry
118 110 16	
119 10 16	is the answer.

So, 106^3 = 1191016.

Example 3 : Find the cube of 98 by Yavadunam Sutra.

Step 1. Nearest base of 98 is 100. Deviation from base = – 2.

1st part = number + (2 × deviation) = 98 + 2 × –2 = 98 + –4 = 94

Step 2. New deviation x original deviation.

= (94 – 100) × –2 = –6 × –2 = 12

Step 3. Cube of deviation = $(-2)^3$ = -8.

Arranging the digits,

94	12	–8	We cannot write -8 in the answer.
94	11	100–8	
94	11	92	is the answer.

So, 98^3 = 941192.

Example 3 : Find the cube of 988 by Yavadunam Sutra.

Step 1. Nearest base is 1000. Deviation from base.

= 988 –1000 = –12

1st part = number + (2 × deviation) = 988 + (2 × –12)

988 + –24 = 964

Step 2. New deviation x original deviation.

– 36 × –12 = 432

Step 3. Cube of deviation = $(-12)^3$ = –1728.

Arranging the numbers,

964	432	–1728	To change – sign of 1728, take 2 × 1000 from 432,
964	430	2000 – 1728	
964	430	272	Since base is 1000, so there should be 3 digits in each digit separator). In this case, there is no carry.

So, $(988)^3$ = 964430272.

Example 5 : Find the cube of 1008 by Yavadunam Sutra.

Step 1. Nearest base is 1000. Excess from base = 8.

1st part = Number + (2 × excess) = 1008 + 2 × 8 = 1024.

Step 2. New excess × original excess.

24 × 8 = 192

Step 3. Cube of excess = 8^3 = 512.

Arranging the numbers,

1024 192 512 Since base is 1000, so there should be 3 digits in each digit separator). In this case, there is no carry.

1024 192 512 is the answer.

So, $(1008)^3$ = 1024192512.

Try these : Find the cubes of the below numbers :

(1) 19 (2) 17

(3) 92 (4) 99

(5) 102 (6) 116

(7) 991 (8) 101

Answers :

1) 6859 2) 4913 3) 778688 4) 970299 5) 1061208

6) 1560896 7) 973242271 8) 1030301

Cube Roots

Cube Roots of Perfect Cubes Using Vedic Mathematics

Cube roots of perfect cubes can be found out faster using Vedic Mathematics.

A perfect cube is the result of multiplying a number three times by itself.

$a \cdot a \cdot a = a^3$

We can also say that perfect cubes are the numbers that have exact cube roots.

It is better to remember the cubes of first 15 natural numbers.

$1^3 = 1$ $\sqrt[3]{1} = 1$

$2^3 = 8$ $\sqrt[3]{8} = 2$

$3^3 = 27$ $\sqrt[3]{27} = 3$

$4^3 = 64$ $\sqrt[3]{64} = 4$

$5^3 = 125$ $\sqrt[3]{125} = 5$

$6^3 = 216$ $\sqrt[3]{216} = 6$

$7^3 = 343$ $\sqrt[3]{343} = 7$

$8^3 = 512$ $\sqrt[3]{512} = 8$

$9^3 = 729$ $\sqrt[3]{729} = 9$

$10^3 = 1{,}000$ $\sqrt[3]{1{,}000} = 10$

$11^3 = 1{,}331$ $\sqrt[3]{1{,}331} = 11$

$12^3 = 1{,}728$ $\sqrt[3]{1{,}728} = 12$

$13^3 = 2{,}197$ $\sqrt[3]{2{,}197} = 13$

$14^3 = 2{,}744$ $\sqrt[3]{2{,}744} = 14$

$15^3 = 3{,}375$ $\sqrt[3]{3{,}375} = 15$

We can summarize the below from the above list of cube roots.

Unit digit of a cube root	Unit digit of a cube
1	1

8	2
7	3
4	4
5	5
6	6
3	7
2	8
9	9

It's very easy to remember the relations given above as follows.

$1 \Rightarrow 1$	same numbers
$8 \Rightarrow 2$	10's complement of 8 is 2 and 8+2=10
$7 \Rightarrow 3$	10's complement of 7 is 3 and 7+3=10
$4 \Rightarrow 4$	same numbers
$5 \Rightarrow 5$	same numbers
$6 \Rightarrow 6$	same numbers
$3 \Rightarrow 7$	10's complement of 3 is 7 and 3+7=10
$2 \Rightarrow 8$	10's complement of 2 is 8 and 2+8=10
$9 \Rightarrow 9$	same numbers
$0 \Rightarrow 0$	same numbers

Also see

$8 \Rightarrow 2$ and $2 \Rightarrow 8$

$7 \Rightarrow 3$ and $3 \Rightarrow 7$

Example 1 : Find the cube root of 512.

Step 1. Make a groups of three digit from right side. Put a bar over the numbers from right to left leaving 2 digits. Number 512 can be written as 512.

Step 2. Since this number has only one bar, cube root of 512 will have only one digit.

Step 3. Unit digit of the number is 2. From the above table it is clear that, if unit digit of the cube is 2, unit digit of the cube root will be 8. Hence one and only one digit of the answer is 8.

So, cube root of 512 is 8.

Example 2 : Find Cube Root of 6859.

Step 1. Make groups of three digits from right side. Put bar over the number from right to left, leaving two digits at a time. So, 6859 can be written as 6 859.

Step 2. Since here bar is placed on two numbers, Cube root will contain only 2 digits.

Step 3. Here unit digit of the number is 9, from the above table we can see that unit digit of the answer will be 9.

Step 4. For ten's digit, next bar is over 6, we know that, $1^3 < 6 < 2^3$. Hence ten's digit is 1.

So, cube root of 6859 is 19.

Example 3 : Find the cube root of 15625.

Step 1. Make groups of three digits from right side. Put a bar over the numbers from right to left leaving 2 digits. Number 15625 can be written as 15 62 5.

Step 2. Since this number has two bars, cube root of 15625 will have two digits.

Step 3. Unit digit of the number is 5. From the above table we can see, if unit digit of the cube is 5, unit digit of the cube root will also be 5.

Step 4. Next bar is over the number 5. Then we have to find a digit whose cube root is less than 15. i.e., $2^3 < 15 < 3^3$. Hence ten's digit is 2.

So, cube root of 15625 is 25.

Example 4 : Find the cube root of 474552.

Step 1. Make groups of three digits from right side. Put a bar over the numbers from right to left leaving 2 digits. Number 474552 can be written as 474 55 2.

Step 2. Since this number has two bars, cube root of 474552 will have two digits.

Step 3. Unit digit of the number is 2. From the above table we can see, if unit digit of the cube is 2, unit digit of the cube root will be 8.

Step 4. Next bar is over the number 4. Then we have to find a digit whose cube root is less than 474. i.e., $7^3 < 474 < 8^3$. Hence ten's digit is 7.

So, cube root of 474552 is 78.

Try these : Find the cube roots of the below numbers:

(1) 17576

(2) 32768

(3) 91125

(4) 125000

(5) 238328

Answers :

1) 26 2) 32 3) 45 4) 50 5) 62